World Architects in their Twenties
建筑师的20岁

东京大学工学部建筑学科　安藤忠雄研究室　编
王　静　王建国　费移山　译

清华大学出版社

Renzo Piano Friday, April ??

Jean Nouvel Thursday, June 18

Ricardo Legorreta Monday, June 29

Frank O. Gehry Thursday, October 29

I.M. Pei Friday, October 30

Dominique Perrault Friday, November 20

东京大学 1998 年建筑与教育讲座系列

伦佐·皮亚诺	4 月 24 日,星期五
让·努维尔	6 月 18 日,星期四
理卡多·雷可瑞塔	6 月 29 日,星期一
雷姆·库哈斯	7 月 18 日,星期六
弗兰克·盖里	10 月 29 日,星期四
贝聿铭	10 月 30 日,星期五
多米尼克·佩罗	11 月 20 日,星期五

目 录

译者弁言 .... 王建国 8

引言

启示和收获 .... 安藤忠雄 11

访谈录

伦佐·皮亚诺 23
RENZO PIANO

让·努维尔 45
JEAN NOUVEL

理卡多·雷可瑞塔 75
RICARDO LEGORRETA

雷姆·库哈斯 97
REM KOOLHAAS
（库哈斯没来）

弗兰克·盖里 100
FRANK O. GEHRY

贝聿铭 127
I.M.PEI

多米尼克·佩罗 147
DOMINIQUE PERRAULT

座谈会

建筑与教育——如何制造"场所" 171

编译者简历

座谈会出席者简历 193

演讲者简历撰写者和注释执笔者简历 ... 195

译者简历 197

责编后记 198

译者弁言

除了像安藤忠雄这样的少数人以外,一位从事建筑设计工作的人在20岁左右,大概应处在大学求学阶段,大学期间的心绪和场景,常常是后来人们追怀反思的有趣话题,也是支持事业前进和学术发展的起点。

建筑学子学习过程中反映出的那种"无知而无畏"的激情和热望,可贵而难得。"年轻"虽然通常意味着稚嫩和脆弱,但也有着对理想的憧憬和富有活力的创新精神。他们对于那些已经成为世界大师级的建筑师心怀景仰,趋之若鹜,乃至表现出如同"追星"般的痴狂,正如安藤所说,"那时我们只要看到世界上有名的建筑师的身影就会感到强烈的震撼,期望着自己有朝一日也能变得像他们那样。"不过对于这些大师早年与他们年龄相仿时的建筑学习经历、所思和所想,今天的建筑学子大多了解甚少。正因如此,安藤假东京大学任教授之际,邀请六位国际建筑大师与学生面对面地交流切磋,其意义是不言而喻的。

近年,中国与国际建筑界往来频繁,学术交流和合作丰富多样,但是就我所知,国外建筑师在高校举行的讲座大多是介绍其已经取得成功的一面,如介绍设计作品、创作理念或国际趋势等,讲与同学们如此接近和具有亲和力的话题还是很少。从这些世界一流的建筑师所谈到的当年经历和建筑理念,我们不难产生不少共鸣，如学建筑的应该多外出旅行以增加阅历；应该多接触建筑实际建造和材料使用以克服当今信息时代与物质时空的疏离；当然更重要的是他们对建筑的情有独钟和挚爱，如果没有这样的信念，当年曾经因经济窘困而开卡车谋生的盖里，或许就不会达到今天这样的世界设计水准和事业颠峰。他们的经历也引发了我们对一些问题的进一步思考，如努维尔和佩罗谈到的美术基本功与建筑创造性之间的取舍矛盾、皮亚诺当年求学时的"半工半读"以及所谓的"无师自通"(No Education)、佩罗完成的别出心裁的毕业设计等，这些都与我们今天常规的建筑教育有所抵牾。建筑教育如何在保证社会对建筑职业人才需求的同时，兼容一定的对多元性和学生自我个性发展的肯定，对这一问题读者自可深思。所以，《建筑师的20岁》不仅是对日本东京大学，对中国的建筑学子同样也有着良好的可读性和借鉴意义。

初识《建筑师的20岁》，我感觉全书字数并不算多，且认为对我国建筑学子有很好的阅读和学习价值，于是承担了翻译的任务。翻译过程中，才发觉翻译此书并非想像的那么简单。其一，因为是访谈性质的图书，书中语言都是口语，口语不如书面用语严谨，句子大多十分简单且随意，这一点给翻译工作带来了较大的困难，个别地方我们只能根据上下文去揣摩建筑大师们的意思，然后再翻译成文。其二，书中内容涉及大量人名、地名和历史事件，对此我们查阅了有关资料和每位建筑师的相关网页，并补充增加了一些译者注，尽量给读者提供更加充实的背景资料，以帮助他们加深对建筑大师们的认识和理解。其三，当时访谈采用的是英语或法语，东京大学工学部建筑学科（相当于中国大学中的建筑学院）安藤忠雄研究室根据访谈录整理了一份日文稿，并加了一些注解，原书出版时则采用了英文／法文和日文对照的形式。我们在翻译过程中，英文访谈部分以英文为基准，日文部分作为参考，这样以保证原汁原味；法文访谈部分以日文部分为基准，同时请我的一位懂法语的研究生沈芊芊进行校对。翻译时，我们尽量做到"信、达、雅"，但难免会有疏漏之处。我们期待读者的批评指正，以期如有机会重印加以纠正。

本书自出版后，受到了广大读者的喜爱，他们给我们提了很多好的建议，例如热心的陈曦明，他在自己的博客中对本书的部分内容提出了很多很多好的建议，我们在此次重印中采纳了一部分，使得本书的翻译更加流畅，生动。在此，我们向这些热心的读者表示诚挚的谢意，是你们使这本书更加完善，也希望大家继续关注本书，多提宝贵意见。

王建国
2007 年 4 月

引言

启示和收获

——由东京大学1998年
建筑与教育系列讲座所想到的

安藤忠雄

世界正在迅速变小。学生们现在能够非常容易地出国，有留学经历的人也不再像以前那样罕见了。建筑师的活动范围在不断扩大，参加国外设计项目的机会也越来越多。

1998年，活跃在世界舞台上的六位建筑师，利用访问日本的机会，从百忙之中抽出时间来到东京大学建筑学院，向我们讲述他们年轻时的事情：他们是怎样对建筑发生兴趣，他们又如何学习建筑，以及刚刚组建事务所时的情况。虽然事先并没有和他们商量过这个系列讲座的事情，但为了我们年轻的学子，这些建筑师都非常爽快地答应了我们的邀请。

那么，为什么要邀请这些建筑师呢？在我们这一代人年轻的时候，根本就没有必要去问这个问题。那时侯我们只要看到世界上有名的建筑师的身影就会感到强烈的震撼，期望自己有朝一日也能变得像他们那样。

但是，最近的学生变得非常冷静。如果要了解某位建筑师，平时通过杂志的介绍就早已非常熟悉，通过CD-ROM更可以随时对其作品进行模拟体验。

在现代社会高度信息化的环境当中，与现实中的人和事进行直接接触所产生的感动越来越少。但是，不管信息多么丰富，不管电脑多么迅速地给我们提供所需要的信息，这些信息无论如何无法与直接的体验相比，尤其是那些亲身感受过生存和创造的艰辛的人面对面向我们讲述，让我们直接从他们那里获取能量和刺激，这实在是一个不可多得的体验。

对于刚刚开始学习建筑的年轻学生而言，他们的谈话对你们产生的直接影响可能不会很多，但活跃在世界第一线的建筑师的存在，他们的气息、动作、话语，以及丰富的表情中折射出的阅历丰富的人生，必定会使大家认真思考自己将来的人生之路，并从中获取生活的力量。究竟从他们的谈话中能够得到些什么，与你们每一位听者的感悟相关。

现在的建筑界正明显地受到信息社会的冲击，建筑设计已从手绘图纸和手工模型的设计方式迅速向CAD这样的虚拟世界转移。而人却仍然要在现实的世界中生存。在20世纪后期，如何解决现代主义与地域主义之间的矛盾是建筑界一个重大的课题，而进入21世纪以后，这对矛盾也许将会被虚拟与现实的矛盾所替代。

在这种状况下，活跃在实际创作中的建筑师们与学生进行直接交流就具有很重要的意义。近距离地听着他们热烈的谈论，感受到他们无法通过媒体传达的能量，这无论对学生还是对我们来讲都是一个非常有刺激性的体验。

十几年前，我在美国的耶鲁大学、哈佛大学和哥伦比亚大学做过一些短期的工作，当时彼得·埃森曼、西萨·佩里、汤姆·梅因、肯尼斯·弗兰姆普顿等人会随意访问我的工作室，就学生的课题进行评论、对话，同时，学生也可旁听他们之间热烈的争论，这实在是一个非常新鲜的体验。他们有一种自然的感情，希望能够将自己的认识传授给下一代的学生，这已成为他们的一个共识，也让我们察觉到无论他们年龄怎样增长，身处什么样的位置，都始终保持了一份自由的精神。对他们而言，直接与学生接触、交谈，与学生共度一段时光是一件快乐的事情。

为了实现建筑的理想，必须克服重重的困难。进入社会开始建筑工作以后，建筑师大部分时间会被杂务所缠绕，从而失去理想，也失去了创造的乐趣和力量。为了不至如此，关键的办法，也是惟一的办法就是看你是否从心底里热爱建筑。我想这样说一点都不过分。这次应邀而来的建筑师尽管年龄和国家不尽相同，建筑风格也迥然不同，但让人强烈感受到的一个共同点就是他们从心底迸发出的对建筑的热爱。

这次参加海外建筑大师系列讲座的第一位是伦佐·皮亚诺。1977年完成的蓬皮杜文化艺术中心❶，可以算是他的成名作。在当年，当这座建筑突然出现在古建筑云集的巴黎街区的时候，它所带来的冲击即使到今天仍令人难以忘却。从此以后，以表现技术为主题的设计思想，也被称为"高技派"的设计思想，成为现代建筑的一大潮流。皮亚诺的建筑，在"高技派"的同时，也具有手工作业的层面，拥有多样性和广阔的内涵，很难用一个单纯的词汇来概括。特别是在大胆而合理的构思中不乏深入的细部设计和多样性的材料应用，成为他的重要特点。从他的代表作梅尼尔收藏品美术馆开始❷，一直到最近完成的拜耶拉基金会美术馆❸，我们都可以看出他的一贯做法，即用单纯的构成进行重复叠加创造出富有韵味的丰富空间，从建筑的整体构图到百叶、窗户等细部处理，这种思想贯穿始终。这种特色也许与他的故乡热那亚海所孕育的温和的人性有关。我与皮亚诺在大阪关西国际机场建设时经常会面❹，但是通过这次讲座，我进一步理解了他的建筑态度以及建筑世家的家庭环境对他投身建筑事业的影响。

❶ Center Pompidou, Paris, France, 1971—1978.
———译者注

❷ Menil Collection Museum, Houston, Texas, USA, 1982.
———译者注

❸ Beyeler Foundation Museum, Basel, Switzerland, 1992—1997.
———译者注

❹ Kansai International Airport Osaka Kansai, Japan, 1988—1994.
———译者注

图中左边第 2 位起依次为：弗兰克·盖里，凯瑟琳·梵德罗，安藤忠雄，汤姆·海奈干

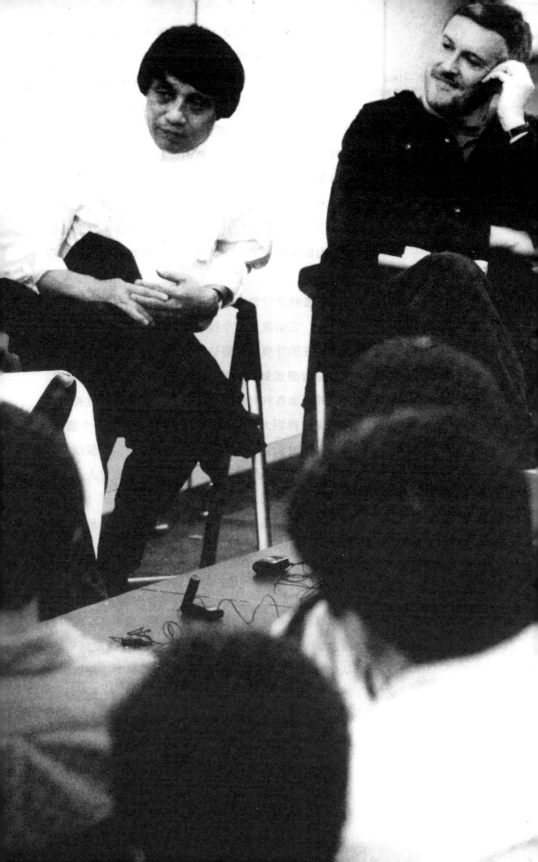

❶ Jean Prouve，1901—1984，法国著名建筑师，工业化建筑的倡导引领者之一．
——译者注

　　从皮亚诺、罗杰斯、福斯特等人的作品中可以看出，他们明显受到先驱者让·普罗维（Jean Prouve）❶的影响。普罗维的特点在于对新技术的执著追求和对工业产品所带来的细部的关注，同时也不放弃手工实践。从皮亚诺的作品中我们可以明显地看到这些特点。同时，在谈到皮亚诺的建筑时，我们也不能忘记阿儒普事务所已故工程师彼得·莱斯先生，他在蓬皮杜文化艺术中心和关西国际机场等工程中为皮亚诺的设计提供了技术上的支持，两人共同建造了许多富有动感和张力的建筑，形成了建筑师与工程师之间的一种理想组合。

　　蓬皮杜文化艺术中心所带来的争论并不局限于建筑形式，它引发了整个社会关于现代美术的走向、过去与现在的冲突融合等许多问题的讨论。站在它的面前，你会感觉到这栋建筑不停地在诉说着这些冲突、矛盾并引出各种对话。如前所述，在能够通过各种各样的媒介得到大量信息的现在，我们总有一个错觉，就是直接对话似乎变得没有必要。但是，不仅仅是建筑界，在建立新的价值观的时候，如果没有人与人之间的交流和冲突，就无法令人感觉到"意志"的存在。在与他人的对话中，各种各样的价值观相碰撞，互相之间给予不同的刺激，从而引发自己的思考，这应该成为创造的原动力。

　　在墨西哥建立了自己独特世界的理卡多·雷可瑞塔，敢于直面现代主义建筑与地域和风土之间的矛盾。从他的建筑中我们可以看出，他的设计方法是以富有活力的水的处理为特征，结合中美洲干燥土地所形成的独特色彩和建筑形态，创造出具有地域特色的建筑形式，从这些朴素的表现中我们能感受到他对建筑创作的热情。在向高度信息化社会过渡的今天，建筑界提出了怎样捕捉地域的特殊性这一问题。路易斯·巴拉甘❷逝世后，雷可瑞塔结合本国文化创造出具有自己个性的建筑创作方法，超越了单纯的建筑形式的问题，这促使我们再次回到设计的起点，反省自己的建筑创作思想。

弗兰克·盖里从年轻时起就与克雷斯·澳典巴克、弗兰克·斯泰拉交往很深，并受到他们的影响。他从建筑与现代艺术的关系中探索建筑设计的可能性，从中获得建筑表现的自由。现在，他在完成被评为20世纪最优秀建筑的"毕尔巴鄂·古根海姆美术馆"❸之后，又完成了在形式上与之相对立的柏林DG银行❹，向世人展示出他进一步的发展。盖里在加利福尼亚工作，这里所具有的自由开放的氛围使他天性中的造型能力得到了充分的发挥。当然，这些成就的背后更隐藏着深层的思考、坚强的意志和艰辛的劳动。从古根海姆美术馆的模型中，我们能看出他从构思到具体实现的过程中所费的苦心，这是仅从建成后的建筑中所无法看到的，这一点有着深层的意义。在这次的讲座中，我们还意外地发现了他还具有战略性思考的一面，这使我们再次认识到他不愧是一位能够在严酷的世界中找到生存之路的人。

让·努维尔以他强劲有力的风格和语速很快的演讲令我们折服。从塞纳河畔的"阿拉伯世界研究所"❺，一直到里昂歌剧院、卡梯尔基金会画廊❻等，他的设计在自由灵活、不断创造新形式的同时，始终保持了高质量的空间和独特的氛围。虽然在每一个作品中都有材料和细部设计的变化，但我们总是能够清晰地感受到努维尔的存在。最近，《建筑文化》杂志（1996年7月、9月、12月）将他未能实现的方案进行了一次汇总，其数量非常惊人。确实，如果没有坚强的毅力恐怕是无法做一名建筑师的。以前我也曾有机会在阿姆

❷ Luis Barragan，1902—1988，墨西哥著名建筑师，地域建筑文化的倡导者。
——译者注

❸ Guggenheim Museum Bilbao，Bilbao,Spain，1997.
——译者注

❹ DG Bank Headquaeterrs，Berlin，Germany，2000.
——译者注

❺ The Arab World Institute，Paris，France,1981—1987.
——译者注

❻ Opera House，Lyon，France，1986,Cattier Foundation Gallery，Paris，France,1994.
——译者注

斯特丹听过他的演讲，那时他也是几个小时不间断一直讲解着自己的作品，当时就令我非常感动。他不停地思考并赋予行动。这一次我们很高兴地看到他顽强的体力和奋发的精神仍然健在，我想学生们定会从中得到巨大的鼓舞。

在我看来，在继承历史文化方面，巴黎是一座拥有独特智慧，并具有深刻意义的现代城市。在这个城市中，多米尼克·佩罗又完成了一个有魅力的建筑，这就是在国际设计竞赛中夺标的法国国家图书馆❶，当这栋建筑以其飒爽的形象建成之后，佩罗又展开了更有活力的活动，最近他又在柏林完成了"奥林匹克游泳馆"❷，作品集也相继问世。在他的作品中，大胆的体量处理、具有金属感与透明感的材料选择给人一种未来建筑的印象，这或许就是他能够吸引对新的表现形式具有敏锐触角的年轻人的原因。佩罗的建筑在夸张性、透明感等方面与让·努维尔有共同之处，这也许是法国人特有的感性。当然，不可否认，这次前来的建筑师都在自己的工作中表露出自己的地域和文化背景。例如佩罗的法国国家图书馆中面向塞纳河的"甲板"部分，平时向市民开放，这与法国人的生活密切相关，在法国，人们习惯于在广场以及室外咖啡馆聚会、交谈，这些习惯可以在这个开放的"甲板"部分得到解决。虽然他是这次的六人中年龄最小的一个，但他说的"我想建造一个场所，而不是建筑。……重要的不是建造什么，而是为什么建造"这句话让我深有同感。

在六人中最为年长的是沿着现代建筑的历程走来的建筑大师贝聿铭。他在第二次世界大战之前从中国来到美国，历尽艰辛后融入了美国社会，他超越了这种巨大的文化差异，始终保持了创造性和灵活的思维，

❶ French National Library, Paris, 1989.
——译者注

❷ Olympic Velodrome and Swimming Pool of Berlin, Germany, 1992.
——译者注

也因此被称为伟大的建筑师。这次的谈话让我感到特别意外的是，直到受密特朗总统之邀在巴黎"卢浮宫美术馆"中设计玻璃金字塔❸之前，他几乎没有思考过建筑设计与历史的关联。在一般情况下，谈到自己的设计思想，建筑师不可能进行这样的表白，这种直率的态度、坦诚的谈话令我深为感动。在当今偏重知识的建筑界，不管是哪一位建筑师都会表现出对任何一个知识领域都非常精通的样子，并像评论家一样展开论述，而这位顶级建筑师却能够将自己的失败在年轻人面前平静地讲出，这中间体现出的丰富的人性，令我受到强烈的感染。贝聿铭讲述了他从年轻时开始喜爱旅行，爱好历史并阅读了大量书籍，从设计卢浮宫的玻璃金字塔开始深刻地思考历史，与历史进行对话。在人们的心目中，他的卢浮宫设计已成为20世纪的伟大遗产。建筑师的职业性格往往使人不可避免地养成注意他人目光的习惯，但已成为大师的贝聿铭所展示出的直率和诚实，令我从心底感到敬佩。

❸ "the Glass Pyramid", or New Entrance to the Louvre, Paris, 1989.
——译者注

对于缺乏外部刺激的学生来说，能够亲耳听到活跃在世界各地的建筑大师的讲述真是一件令人非常喜悦和非常感动的事情，这不仅对学生，对我本人亦是如此。

建筑是一种揭示理念、构筑新的"价值"的行为，在这个过程中必须克服物质的、经济的以及社会的各种障碍。如果说光临我们大学的六位建筑师中有什么共同之处的话，那就是他们都具有坚定的信念和将其贯彻到底的坚强意志，以及能够冷静而客观地剖析自己的特点。为了"自由"和更多的可能性，他们不断地与自己和社会进行着斗争。

从这次的系列讲座中学生们能够得到什么呢？对于为了21世纪社会而学习的学生们而言，这些优秀的建筑大师们充满人性的话语，和他们殷切的期望，一定会深深地感动我们，并注入我们莘莘学子的心田。

现在对你们而言是人生中最关键的时刻，如果这些大师"认真生活"的态度和维护自由尊严的精神，能够成为今天我们再一次思考的契机，则是再好不过的事情。在他们看来，"年轻"虽然具有脆弱的一面，但也有着勇往直前的坚强力量。

安藤忠雄

建筑师。1941年生于大阪。自学建筑，1969年成立了安藤忠雄建筑事务所。

历任耶鲁大学、哥伦比亚大学、哈佛大学客座教授。从1997年11月开始任东京大学教授。

代表作有"兵库县儿童博物馆"（1989）、"光的教堂"（1989）、"1992年塞维利亚世界博览会日本馆"（1992）、"大阪府立近代飞鸟博物馆"（1994）等。在1997年的国际建筑设计竞赛中入选的"福特沃斯现代艺术博物馆"、"兵库县立新美术馆"等项目现在正在进行中。

他自日本建筑学会奖（1979）得奖以来，获得了难以尽数的国内外奖项：法国建筑奥斯卡大奖(1989)、日本艺术院奖(1993)、1995年度普利策奖（1995）、高松宫殿下纪念世界文化奖(1996)、英国皇家建筑师协会（RIBA）金牌奖等。

1995年1月阪神·淡路大地震以来，作为复兴支援十年委员会执行委员长尽力开展灾区的复兴工作，并开始从城市规划的角度开展建筑和建筑师的工作，将活动的范围进行了更大的扩展。

访谈录

刚刚独立时的皮亚诺,在最初的热那亚事务所中(1969年)

1998年4月24日（星期五）10:30am — 12:00am　主持：大野秀敏

伦佐·皮亚诺
RENZO PIANO

1937年生于意大利的热那亚。1964年毕业于米兰技术学院。从大学在校时开始从师于建筑师弗兰克·阿尔比尼。1971年成立皮亚诺 & 罗杰斯事务所、1977年成立皮亚诺 & 莱斯事务所。现在热那亚、巴黎、柏林设有伦佐·皮亚诺建筑工作室，进行设计活动。

在1971年的设计竞赛中被选为"蓬皮杜文化艺术中心"的设计者（与理查德·罗杰斯合作）。当时他只有33岁。1977年"蓬皮杜文化艺术中心"完成。在这栋建筑中，涂成原色的钢铁桁架、通风管道、电梯井道、圆形自动扶梯通道覆盖了整个外立面，这个外观造型给当时建筑界极大的冲击。

人们初次看到"蓬皮杜文化艺术中心"的设计会认为这是异想天开，但是，在这个建筑背后，我们能隐约看到皮亚诺多面而综合的建筑师形象——一个与达·芬奇、米开朗琪罗共同的文艺复兴时期万能人的形象。

皮亚诺具有明快的空间构成能力，找出技术问题解决方法的科学家的分析能力，将理性的设计方案升华为空间美学的艺术家的构想力，融合技术与美学的细部处理，并将其应用于建筑设计之中的工匠的技术能力，除此之外，他还能将环境的潜在特性外显成城市设计策略。

能够反映出皮亚诺多面性和综合性特点的第一个作品就是"蓬皮杜文化艺术中心"。在这之后近30年的设计历程中，他的这种特点会出现在每一个作品中，但每次也都会在思路上有所偏转，并形成新的特色。

（岩城和哉）

新喀里多尼亚的奇巴欧文化中心/1998 © John Golings

蓬皮杜文化艺术中心/1977 © K.Iwaki

无　师　自　通

❶ 1937年正好是勒·柯布西埃诞生（1887年）的半个世纪以后。同年出生的建筑师还有拉菲尔·莫奈（Rafael Moneo），谷口吉生，香山寿夫等人。

❷ 冈部宪明，1947年生，早稻田大学毕业后，1974年留学法国。1974—1977年在皮亚诺＆罗杰斯事务所工作。1978年成为皮亚诺巴黎事务所的设计助理。1988年为了关西机场的建设项目担任大阪事务所的代表。1995年成立冈部宪明建筑师网络。现在是神户艺术工科大学教授，东京大学兼职讲师。另外，在东京大学研究生院开设的设计课题的题目是"机场"。

❸ 日本建筑学会奖：日本建筑学会每年对日本国内的作品进行评选，选出两三名建筑师授予此奖项。在过去的获奖名单中可以看到许多日本著名的建筑师。由于评选对象是日本国内建筑作品，有时也会有外国建筑师获得此奖。曾获奖的外国人有伦佐·皮亚诺，雷姆·库哈斯等。

❹ 普利策奖：也称为建筑界的诺贝尔奖，是世界上最权威的奖项之一。在过去的获奖者名单中可以看到许多世界上有代表性的建筑师。

大野——伦佐·皮亚诺先生1937年出生于意大利的热那亚❶，1964年毕业于米兰技术学院。通过和理查德·罗杰斯先生合作设计"蓬皮杜文化艺术中心"而在世界一举成名，并获得许多奖项。他在日本和冈部宪明❷先生合作设计的"关西国际机场"获得了日本建筑学会奖❸和高松宫殿下纪念世界文化奖。今年（1998年）又获得了被称为建筑界的诺贝尔奖的普利策奖❹。

安藤——今天我们有许多问题要问伦佐·皮亚诺先生，作为学习建筑的学生，我们想请皮亚诺先生谈一下他所接受的建筑教育。

大野——您接受的是哪一种建筑教育？

皮亚诺——我，没受教育（笑）。

　　言归正传，我觉得今天应该是一次坦率直白的对话，所以我应该都说真实的东西。事实上，由于我出生于一个建筑商的家庭，所以我从孩提开始就决定要成为一名建筑师，这有些像出生于杂技世家的人。如果你生于杂技演员的家庭，你成为一个杂技演员就是一件顺理成章的事，而如果你生于建筑商的家庭，你

HIDETOSHI OHNO:
What kind of education have you received?
RENZO PIANO:
Me? No education (laughing).
Well, this is a conversation a 'candid' conversation, I guess, so you have to tell the truth. And the truth is that I decided to become an architect when I was a child, because I was growing up in a family of builders and this is a bit like growing up in a family of acrobats. And you don't wonder what you do if you're in a family of acrobats; you become an acrobat. If you are in a family of builders, you become a builder. Of course, being an architect is not the same as being a builder. It's actually as my father had once told me, "less than being a builder." Being an architect is worse than being a builder. That was my father's opinion. But still, I wanted to be an architect. I do not know why. I was very obedient as a child. As a young man I was not really obedient; I was always doing the opposite of what I was told. So I wanted to become an architect.
I also felt that being an architect would be more creative, would be more free than be-

成为一个建筑商也就是一件顺理成章的事了。当然，一位建筑师与一位建筑商是不同的，我父亲曾经告诉我："不要从事比建筑商更差的职业。"当建筑师要比当建筑商糟糕，这就是我父亲的观点，但我仍然要当一位建筑师，也不知道为什么。其实，我孩提时是很听话的，不过到了青年时代就不一样了，我总是做与我被教诲所相反的事情，于是我要成为建筑师。

我觉得，相比建筑商而言，建筑师（的工作）要更具有创造性，也更加自由，这就是我想当建筑师的理由。某种意义上讲，这就是我的最初的建筑教育。建筑对我来说是一个神奇的东西，因为你可以将一些砂子、石头和砖块这样的东西转变成一幢建筑，这就是神奇。

我必须坦率地承认，这也是我进大学学习的理由。（笑）虽然我进大学了，但我并没有呆在大学里，而是一直在当时米兰一位知名建筑师弗兰克·阿尔比尼❶的事务所工作，我在该事务所没日没夜地工作，从不到学校上课。现在我必须停顿一下，我并不希望在座各位认为大学是无用的，上大学很好，你们必须上大学！

但我要说的是，那时的大学不如今天的大学这么好。当时欧洲一些年轻学生在大学造反并试图在政治上统治学校，米兰技术学院也如此，院系则是最先被

❶ 弗兰克·阿尔比尼（Franco Albini，1905—1977），意大利建筑师，1925年毕业于米兰技术学院。

ing a builder. And so that's why I decided to be an architect, but you know in some way, this was the first education; building for me was really something like magic. You transform things, like sand, stone, bricks into a building. Now, this is magic!
That's why I went to the university, I have to be honest: I went to the university, but I was never at the university because I was working with Franco Albini who was a great architect in Milan at that time. So I was working from morning to evening in Franco Albini's office and I was never going to the university. Now I stop because I don't want you to believe that university is not good. University is very good! You must go to the university!
But, at that time, the university was not as good as it is today. I think it was very interesting but I was going to the university in the evening because that was the time in Europe when young students began to invade university; to politically occupy schools. And so in Polytechnic of Milan, the first place where university was occupied, was the faculty. So I was normally working the day

占领的地方，因而我都是晚上去学校学习，我认为这很有趣。我通常白天工作，晚上回到学校，当然是回去睡觉。（笑）

建筑就像浮在大海里的冰山

大野——我比皮亚诺先生晚一两年进入大学，日本也有大学被学生占领的时代。虽然当时皮亚诺先生只在夜间去学校，但我想在那里也还是会得到各方面的信息和影响。您能谈谈这些大学经历对你后来的专业生涯有什么影响吗？

皮亚诺——上大学对我是非常基本的，因为正是在那个年龄段我开始感到了一种现在所谓的"了解社会的渴望"。这是一种对社会的好奇心，一种知识上的贪婪。而这对于成为建筑师是很重要的。仅仅在设计、构思空间、体量和形式，或者在科学、"科学家"方面做到完美是不够的，但是具有社会愉悦感与社会责任感对于建筑师来说是很重要的。

1962年至1964年间是我建立个人经验的阶段，我将其称之为"知识好奇"和"社会关注"时期。这是非常重要的，正如你们所知，我经常讲"建筑是一座冰

and then in the night I was going to the university to sleep, of course. (laughing)
HIDETOSHI OHNO:
Can you describe how you feel you have been affected by your experiences of university days to your professional career afterward?
RENZO PIANO:
It was essential for me to go to the university, because at that age I began to feel what I now call "anxiety about sociality." It's a kind of social curiosity, an intellectual voracity. And this is essential to being an architect. It is not enough to just be "good" at designing, thinking space, volume and form, or being very good at science, "a man of science", but also it's also very important to be socially comfortable, socially motivated.
And those years between '62-'64, between those years for me was the time when I built up a lifetime of personal experiences which I call "intellectual curiosity" and "social interest." And this is essential, because you know, I keep saying, "Architecture is like an iceberg," and the visible part is very little, but all of the part below is the one that makes architecture, and the part below is society,

山",其真正可见的是浮在水面上的很少一部分,而水面下的才是建筑的主要部分,其中包含了社会、人类学、历史、地理学、气象学、科学和社会科学等,缺少这些部分的支持,建筑是不存在的。所以,这是纯粹的学术问题。

就像孩子在海边玩沙子一样,我的青年时代是在玩弄材料

大野——看到伦佐·皮亚诺先生的作品之后,我们深受感动的原因之一,就是工匠的手艺与现代技术完美地结合在了一起。关于工匠的手艺这方面,我们可以从刚才讲到的皮亚诺先生的家谱中看到一些渊源,但是关于现代科学技术这一方面,您是在大学以外的什么地方学习或者研究的呢?您能不能给我们谈谈这方面的事情。

皮亚诺——你们知道,当一个人到了60岁时,要准确记忆所发生过的一些事是不容易的。在我职业生涯的前10年——也可能少于六七年——是没怎么做建筑设计的❶。我像沙滩上的小孩一样玩耍。我那时一直在玩,非常天真无知。我现在仍旧努力保持那种天真,尽管我当年确实无知。

说实话,我的兴趣非常有限。但我确实对材料及材料摆弄非常有兴趣。这种挑战很简单,用很有限的

❶ 1966—1970年皮亚诺的作品有:"皮亚诺办公室"(热那亚)、"大阪世博会意大利馆"(大阪)等。另外,在20世纪60年代末和路易斯·康合作设计了"奥林维蒂·安德武德工厂"(Roofing Components for the Olivetti-Underwood Factory,宾夕法尼亚州)。

anthropology, history, geography, climatology, science, sociology, All these are there. Without that part pushing up, architecture doesn't exist. So, it's pure academism. You know, when you are in your sixties, it's difficult to remember exactly what happened. But, I think that for the first 10 years-maybe less than 6 or 7 years-of my professional life, I didn't make architecture. I was playing like a child in the sand of the beach. I was playing, playing, and I was very innocent. I still try to be innocent, but I was certainly very innocent at that time.

And my interest was quite limited, to be honest, quite limited. I was really very interested in materials, and playing with materials. The challenge was very simple: to do buildings with very little material. It was about lightness, it was about 'immateriality.' It was not very complex. It was probably coming from a sort of challenge, because I had been growing up in a family where buildings were made by concrete, by heavy things. So I wanted to do the opposite, probably.
Also, you know I was very familiar with harbors. I was born in Pegli, a suburban town

材料盖房子,追求轻盈性和非物质性的确不是非常复杂。这或许是来自一种挑战,因为我成长在一个用像混凝土这样的厚重建筑材料盖房子的家庭,所以我想反其道而行之。

你们知道,我对港口很熟悉,我出生在热那亚[1]城郊小镇佩里,并成长于热那亚。热那亚拥有地中海最大的港口,景观非常壮观,这是一种非常轻盈、日复一日不断变幻(转瞬即逝)的场景:水中的倒影,各种反射,穿梭往来的船只。这种轻盈感、可变性和"短暂性"对我非常重要,我也是以此作为起点,说来非常简单,后来我才开始以某种方式去探索复杂性。

当然了,我可不希望自己比表现得更加天真,这只是开始!之后,才开始学习,因为建筑并不是玩耍砂器,而是远远超出其内容。事实上,我花了至少50年的时间才成长为一个建筑师,这确实是一段漫长的(人生)旅程,很漫长。

我是作为一个建筑商的儿子开始的,就像每个人都会以某种方式起步一样。就是这一起步花费了很长时间。直到现在,我看上去才基本上像是一个建筑师了。这花了相当长的时间,因为你必须"积累"(你的知识)。

我认为我必须告诉你们,因为你们现在正在建筑

[1] 热那亚:建筑师往往会有很多机会在自己的家乡进行设计,皮亚诺也是这样,他在家乡热那亚有很多作品:"地铁车站"(1991),"哥伦布发现新大陆五百周年纪念博览会"(1992),"联合国教科文组织研究所及工作室"(1991), Headquarter Harbour Authorities (1995), Completion open spaces, Old Harbour (1996—)等。

of Genoa and grew up in Genoa. Genoa has a big harbor, the biggest in the Mediterranean Sea. And the 'paysage'-the harbor landscapesis fantastic! It's very light, it's very ephemeral-it changes all day: you have the water that doubles the image; you have reflection; you have ships coming and going, 'changing of a part.' So this idea of lightness and the changeability, "temperament, " was for me, very important. So I started that way. Very, very simple. And then I started to draw up in complexity in some way.
Of course, I don't want to look more candid than I am! This was the beginning! Then, you start to learn, because architecture is not just 'playing with sands.' It's more than that, much more. So then I look, but you know it takes 50 years to become an architect-at least. It's a long, long, long travel. It's a long journey.
I started as a son of a builder, because everybody starts some way. But it was one way to start, and then it took a long time. And now, I 'start' to become almost like an architect, you know. But it takes a long time because you have to 'add' things.

系学习，处在一个需要"属于自己的东西"的开始时刻。我的意思是说，在这种时候，"玩耍砂器"只是一个实验，这太傻了。后来，当我慢慢长大了，我就开始思考，这（玩耍砂器）对我而言很好，因为这是我自己的方式，一种自发的方式。当你们年轻时，你们应该看到这一点。我意思是说，我成长于一个居住在热那亚中部地区的建筑商的家庭，以及热那亚这样一个港口城市，而你们每个人都有自己不同的家庭、出生地或来自于不同的地方。每个人都有自己的根。我觉得，你们不要低估你们曾经拥有并可深度挖掘的个人经验的重要性，因为这是你们每个人的（成长）历史，也就是维系你们自己的根脉。有人曾经说，一个人在孩提时代，其习性实际上就已定格，而在他后来的人生中所做的只是对其经历的追根溯源而已。我认为，一个人不能忘记你来自何处、你的出生地和那些属于你自己而别人所没有的东西，理解这一点非常重要。换句话说，不要像我这样，也不要像安藤或是大野那样，只要像你们自己，不要以别人作为自己的样板，你们必须构筑你们自己。

孩童时期

© Renzo Piano Building Workshop

But what I think you should be told, because we're in the school of architecture, is that at the beginning you need this 'something that belongs to you.' I mean, at the beginning, I thought that this 'playing with sand' was only an experiment, was quite silly. Then, when I became older I started to think that that was very good because that was 'my' approach, my autonomous approach. And when you are young, you have to watch this. I mean, I was born with family of builders dig-in the middle of Genoa, a harbor city, okay? But everybody [you] has been. born somewhere, from some family, from some place. Everybody has roots. I think that you should not underestimate the importance of keeping what you are and then dig-in depth for it, because your personal history is like your own roots. Somebody once said, that in childhood-when you're small-you have really done everything. You then spend the rest of your life digging back into what you've been when you were a child. So I think that it's essential for everybody to understand that you should not forget where you come from,

越是在年轻的时候，
越是要努力培养自我决断的能力

❶ 作品集:《航海日志》(伦佐·皮亚诺，石井俊二监制，田丸共美子等译，TOTO出版，1998)。这是为了配合世界巡回展的一部由皮亚诺本人编辑的作品集。日语版于1998年4月25日（本次讲座的第二天）至6月27日之间在"间画廊"展览期间同时出版。

大野——在TOTO出版社出版的作品集❶的序文中，皮亚诺先生谈到了现代建筑运动面临的困难。刚才您谈到您是怎样坚持自己的志向，用了50年的时间成为建筑师的。

您刚才的谈话给了我们非常大的鼓励。今天在座的学生大部分都是刚刚进入大学一两年，才开始学习建筑，他们从现在开始要考虑就业等问题，面临人生重大的决断，对于这些学生来讲，在50年这样漫长的道路中，最初的两三年意味着什么？您能不能给我们谈一谈这方面的看法。

皮亚诺——当然，这不容易回答，因为我坚持认为，建筑是一门非常非常复杂的艺术。我常说："建筑是一门被生活所感染的艺术。"说真的，这虽然是负面消极的，但同时也是正面积极的，因为这是真实的。建筑艺术受到如此的感染，以致人们无法将建筑与生活区分开来，可以说这是极为普遍的认识。我觉得，对于年轻建筑师而言，发挥自主性以及在做决定时保持自我、充分自信和充满好奇心是很重要的一件事。不要

where you were born, and what things belong to you, not to somebody else. In other words, don't do like me. Don't do like Mr. Ando, don't do like Mr. Ohno. Do like you! Don't take the example of anybody. You have to build your own autonomy.

HIDETOSHI OHNO:

You mentioned the difficulty of keeping oneself develop to be an architect. Would you give some advice for the students who have just started to learn architecture?

RENZO PIANO:

Well, it's not easy, of course, because architecture is very very complex art-I keep saying, "It is an art contaminated by life." And this is negative but it is also positive, because it's true-this is very true. Art is so much contaminated that you cannot separate architecture from life. Anyway, even said so it's a very general consideration, I think the essential thing-for young architects-is to be autonomous, to be free to make decisions…to be 'competent' enough and curious enough to make decisions. Don't believe people, don't believe anybody. Don't believe me, either. I think you have to understand 'yourself.'

相信别人，不要相信任何人，也不要相信我。

你们一定要理解你们自己。自主性是很重要的，遗憾的是这并不容易做到，因为你们必须具有两种重要的品质：一是对生活、对一切事物的好奇心，不仅只是对你们的朋友，而要对"生活"和一切的事情好奇。

另一种品质是，你必须足够地坚韧不拔。你必须这样！你必须知道如何去做事情。否则人们会告诉你："这是不可能的！这是不可能的！这是不可能的！"所以，要做出自己的决定，要会说："不，我要这样做，不要那样做！"而这只有在你足够自信时才能够做到。

人们说技术是重要的，并不是因为技术本身重要，而是因为赋予了你做决定的自由度。所以，当我

演讲会现场
会场在制图教室，学生们围绕皮亚诺而坐。

Autonomy is fundamental, and unfortunately it's not easy, because to be autonomous you have to have two important qualities: one is to be curious about life, about everything, not curious only about your boyfriend, but also curious about 'life,' about everything. Because this is what is giving.

Another thing is that you have to be bloody competent. You have to be bloody good! You have to know how to do things. Otherwise, people will tell you, "You know, this is impossible! This is impossible! This is impossible!" So, to be free to make a decision, to say "No, I do it this way and not that way!" is only possible if you are competent.

People say technique is important, not because technique is important but because it gives you the freedom to make decisions. So when we talk about 'autonomy', it's not just a state of the spirit. It's more than that. It means that you have to be good technically and scientifically, and autonomous enough to support your decision, and the other thing is that you have to be curious, I mean, curious about the sense of life. Because in some way you have to be a humanist. You have to be a real

们谈到"自主性"时,自主性并不仅仅是一种精神状态,自主性要超越它。我的意思是,你必须具备良好的对技术和科学属性的认识,以及足以支持你想法的自主性。此外,你必须具有对其他事物的好奇心,我的意思是,对于生活的好奇心和求知欲。从某种程度看,你必须是一个人文主义者,而且必须是一个真正的人文主义者。你必须具备一个伟大的建筑师的品质,同时还应具备人文关怀的品质。

自主性的重要在于它是一种自由,一种创造性的自由。我觉得你们在大学里可以培育这种自主性。因为在大学里,有老师教你们怎么进步。你们不要停滞不前,你们应该自己走,坚持一些东西,并对一切事物好奇。但这还是不够的。要想获得自由必须有一种强烈的个性、强烈的自主性,同时,你们还应该具有我所有的另一种基本品质,这就是"顽强",一种真正(纯粹)的顽强。

做到顽强并不容易,但是这很重要。这必须是一种"超越性"的顽强。必须是但又不仅仅是这样(掌声)。如果缺乏这种"超越"意义上的顽强,你就只会停留在艺术的边缘。你并没有真正"进入"艺术,而且是停留在表面。顽强是不够的,因为你也许会做错事。这不是顽强,而只

humanist. You have to posses the qualities of the great builder but at the same time the qualities of the culturally minded people.

Autonomy is important, because autonomy is freedom. Freedom. It's creative freedom. So I think in the university you can build this autonomy. Because in the university there are people that may teach you, just go through things. You don't stay in circles. You have to go and you have to insist, you have to be curious about everything. But unfortunately, it's not enough. To be free to have the freedom, then you have a strong character, strong autonomy, but it's not enough, because then you need another big quality that I have, because I have built that quality. And this is 'obstination,' just pure obstination.

Obstination is not easy but it is essential, it must be a 'subliming' obstination. Must be not just…[smack!] like that. Must be 'sublime' in the sense that without real sublime obstination you stay at the perimeter of art. You don't get 'in' art. You stay around, but you don't get 'in'. Subliming obstination, for me, doesn't mean that you decide some-

能说是愚蠢。

所以，说真的，顽强必须与另一个更重要的品质联系在一起，这就是"启蒙智慧"。你必须具有某种吸纳能力，你应该既有智慧而同时又具有吸纳能力。你必须聆听别人所言，这是非常重要的，因为这是矛盾的。一个人只是固执己见很容易。你不用听任何人所言，你只是做你想做的事情。但这不好，事实上也很无趣。

你必须顽强而同时坚韧，具有吸纳能力。这种智慧必须是坚韧的，必须建立在你聆听他人和了解世间一切事物的基础上。

一天中要有一段时间
一个人静静地度过

大野——我完全同意你讲的这些。不过，现代社会媒体非常发达，在我们周围有各种各样的信息。学生们虽然也想自己进行判断，但有时可能会受到杂志的影响，有时虽然想听这个人的话但又受到另一个人的影响，思想容易摆动，非常不容易进入你所讲的状态。

皮亚诺先生在他们这个年龄的时候，也有一个价值观形成的过程，我想也会有许多迷惘，你是怎样形

thing and you just do it. This is not enough because you may do the wrong thing. This is not obstination; this is just being silly. So to be a true, obstination must be copulated with another bigger quality, that is, 'enlightened intelligence.' So you have to be permeable. You have to be intelligent but permeable. You must listen to people. You must listen. And this is very important because it's almost a contradiction. Because it's quite easy to be obstinated if you are just obstinate. You don't listen to anybody. You just do what you want. Not very good. This is not really interesting.
You have to be obstinate but at the same time, light, permeable. The intelligence must be light, must be done in such way that you listen to people, you listen to everything.
HIDETOSHI OHNO:
How did you make up yourself in your youth?
RENZO PIANO:
There are too many magazines, too much of information. In my time, I only got one magazine every two months, and I didn't read it. This is quite true for everybody, not just for architecture. There's too much

成自己坚强的意志的呢？如果你能给我们讲一些这方面的轶闻，我想一定会给我们的学生带来勇气。

皮亚诺——（现在）有太多的杂志，太多的信息。在我年轻时，每两个月的时间才能得到一本杂志，而我并没有读它。这对每个人来说都可能发生，并非只针对建筑师。现在有太多的信息环绕，没有时间去好好思考。就像某个人吃得太多没有时间去消化一样！这对任何事情都是一样的，我的意思是，在我们的时代有着太多的嘈杂喧嚣！太多了！每个人"知道"的越来越多，但同时理解的却越来越少。

我认为太多的信息就像毒品一样，慢慢你会因为毒品而形成依赖性。这是非常坏的毒品，因为你的需求量会越来越大，最后你会不由自主地停止思考。所以我有一个建议：不要买太多的杂志——建筑类杂志。不过，这本书是个例外，否则《皮亚诺作品集》（TOTO出版）的编辑会不高兴的，那就买一本吧。（笑）

即使你们问我这样的案例，也是不容易回答的，因为我有我的戒心，我的意思是说，每个人都有一种戒备意识。我出生在意大利热那亚一座宁静的小镇，作为一个青春焕发的年轻人我经常漫步海滩观海泛舟。我在青年时代敢说敢做，但我也用大海创造宁静。我必须指出，在这一点上，关于宁静和内在性有着深

information going around and there's not enough time to think it off. It's like somebody eating too much. Then, there's no time to metabolize! And this is true for everything. I mean, in our time there's too much noise! Too much! Everybody 'knows' more and more and 'understands' less and less.
I think too much of information's like a drug. It's like a drug, and then you become dependent. It's a very bad drug, because then you need more and more and more and more, then you stop thinking, in an autonomous way. So I have a suggestion: don't buy any more magazines-architecture magazines. Well, I mean, I shouldn't say don't buy 'that' book. I think the editor of "The Logbook, TOTO Shuppan" won't be very happy. Buy just one copy. (laughing) Even though you asked me for an example of this, it's not easy to answer because how I built up my defense-I mean, everybody has a defense. I was born in the quiet, small Italian city of Genoa, and I was sort of adolescent, a young guy walking on the beach looking at the sea. I mean, that's what 'braveur', I used the sea, when I was

厚的文化传统。我认为对于一个人来说，构筑某种戒心，一天中某个私密宁静的时刻，一个你只是安静端坐的时刻应该是可能的。我通常一边安静端坐，一边抽着我自己做的小雪茄。不要去抽烟（笑），但可以像我常常做的那样去静思冥想，例如，我在巴黎有一个办公场所❶，一个主要的办公所在地，它位于城市的中心，我在那儿的工作非常繁忙。同时，我在热那亚也有一个办公室❷，那是在一座山上，没有街道可以通达那里，人们只能使用小型电梯，就像一座孤岛一样，在冥冥之中的岛，而这在今天已经成为可能，因为你可以通过电子邮件、通讯网卡、计算机与世界各地的客户保持联系，即便是你处在一个非常隐密的办公室之中。

所以，我会在余生，像我年轻时，就像你们现在那样，把我的私密性防卫到"宁静时刻"。这不仅仅是欧洲文化，其实更像日本文化，"宁静时刻"是一种反思和内省，我想你们应保持它，由于信息像炸弹一样，所以要做到这一点是不容易的。这就像一种全天候的永不停息的河流。你们应该不时保持清醒，去冥想，否则就会吃得太多不消化了。

❶ 巴黎事务所：在蓬皮杜文化艺术中心附近（距蓬皮杜文化艺术中心大约100米左右的地方），由于"IRCAM大厦"（1977）及其大厦增建工程（1990）、蓬皮杜文化艺术中心周边地区改造方案（2000）等项目，皮亚诺在蓬皮杜文化艺术中心完成后仍一直与该地区保持联系。

❷ 热那亚事务所：即"联合国教科文组织事务所及工作室"，也是皮亚诺自己的作品。

very young, to silence. But I must say that in this character there's great cultural tradition about silence, about interiority. I think it must be possible for everybody to build up the defense, a very private moment in the day-a moment of silence, a moment when you just sit down quietly. I normally sit down quietly and smoke my small cigar. Don't smoke the cigar (laughing), but just sit quietly and try to understand things, as I often do. I mean, I have an office in Paris for example, that is my main office; it is more busy, it is in the center of the city. But I also have an office in Genoa that is on a hill. I don't even have a street to go up there. You use a small lift, and it is like an island, it's isolated, in the middle of nowhere. This is, by the way, possible today, because by e-mail, modem, and computer technology you may communicate with a client around the world even if you are in a very, how do you say, private situation?

So, I was like that when I was young, when I was like you, and then I spent my life defending the privacy as some 'moments of

热那亚事务所「联合国教科文事务所及工作室,第一排中间是皮亚诺,右数第二人是冈部宪明,第四人是石田俊二。

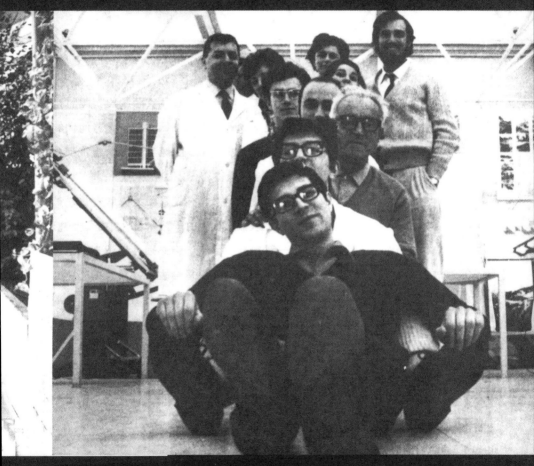

刚成立的热那亚事务所 © Renzo Piano Building Workshop

与安藤忠雄的相遇是一件糟糕的事情，但却从此有了美好的交往

学生——刚才您谈了成为建筑师的五十年的历程。我想问一下在您的诸多经历当中，与安藤老师的相遇是怎样的一个经历？

皮亚诺——唉！真糟糕！（笑）

当你作为一个建筑师工作时，肯定会有一些朋友，也会遇到朋友。我承认，我从安藤那里，同时也从其他朋友那里学到许多东西，因为我认为，他们的品质对于建筑来说是重要的。而像我这样的建筑师几乎就是一个掠夺者（强盗）。艺术就像抢劫一样，从真正意义上讲艺术是抢劫，这不仅是因为你从安藤或其他人那里取得了教益，而且由于艺术与"贪婪"相关，所以这还是一种抢劫。因此，你必须去认识人，也必须去索取一些东西。人们总是会去寻找某些东西，如从安藤已经和正在做的事情中去寻找某些东西。我的意思是说，三宅一生❶正在这里：我就会想他几天前送给我的一本书——埃尔文·宾❷的摄影作品集。你观赏这些照片，同时就会从中抢劫到一些东西。所以，我的整个一生都在索取东西，当然也付出一些，但总的

❶ 三宅一生（1938—）：服装设计师，广岛县出生。1964年多摩美术大学毕业，后赴法深造。在巴黎和纽约工作后，于1970年在东京成立设计工作室。在讲座的当天，三宅一生来到制图教室，和学生们一起听了皮亚诺的讲座。讲座结束后，一部分学生不是请求皮亚诺的签名，而是跑到三宅那里要求签名。

❷ 埃尔文·宾（Irving Penn）：摄影师。其代表作是将为Vouge杂志拍摄的照片汇编而成的作品集Moment Preserved(1960)等。

silence'. And this is not only European culture. This is even more Japanese culture, but 'a moment of silence,' is introspection. And I think you are to save that. It may be difficult, because information is like a bomb. It's kind of an all-day, pandemic river-never-stops. You have to stay awake, from time to time, and you just think about it 'in silence.' Otherwise, it's like eating too much. It's like eating too much, and then you have no time to metabolize.

STUDENT:
How was the first contact with Prof. Ando?

RENZO PIANO:
Well, it was terrible! When you work as an architect, it's essential to have friends. You meet friends. And I learned a lot from Ando, I have to say, and from other friends because I think their qualities are essential to architecture. And an architect is, for me, a robber! Art is like robbery. Art is robbery. Art is robbery in a real sense, not because you take a beeper from Ando or you take a beeper from somebody else, but it is robbery because art is about 'voracity'. So you have to know people and you have to take things.

来说是索取。

在关西国际机场❸工程施工期间，我在大阪的办公室离安藤的办公室不远，所以我与安藤经常碰面。我们一边吃着美味的寿司一边谈着各种话题，但我们一般不谈建筑。我想之所以会是这样，首先是因为他不谈！（笑）其次，这并不是很有趣。你谈论建筑只能说你所做的事情，说你只是盖房子的就够了。有安藤以及其他朋友很好，因为不会使你感到孤独，你总是生活在一定圈子里。由于建筑需要一种合作，所以创造性是非常重要的。这很好，即使是很糟也很好！

❸ 关西国际机场：建于日本大阪湾中15平方千米的人工岛上。1988年，通过设计竞赛选出皮亚诺工作室进行设计。1994年完工。

在巴黎的传统中软着陆的"蓬皮杜文化艺术中心"

学生——建筑师必须是一位能够将建筑层面的东西和文化层面的东西相融合的人文主义者，为什么在"蓬皮杜文化艺术中心"❹的设计中，您能够在巴黎这样一个传统文化城市中大胆地插入一个现代建筑？另外，我们也想知道生于意大利，活跃在巴黎的皮亚诺先生的文化背景。

皮亚诺——你们知道，热爱传统并不等于复制传统。这是非常重要的，因为那时注重的会是美的东西。日本

❹ 蓬皮杜文化艺术中心（巴黎，1977）：一个展示美术、音乐、工业设计的综合设施。总建筑面积10万平方米，平均日访问人数为2.5万人。建后10年中来访总人数为1.5亿人。1971年，在681份参选方案中，皮亚诺和罗杰斯合作的设计方案中选。该建筑的最大特点是将通常需要隐藏的结构、管道、电梯、自动扶梯等部位涂上原色，展示在建筑外部。同时，由于将这些部位放在了建筑外部，建筑内部各层形成了完整的无柱空间。

They are always looking for something, you know, looking for what Ando has been doing. I mean, Issey Miyake's here: I was looking at the book he sent to me the other day-a beautiful photographs by Irving Penn. You look at those photographs and you may rob things from them; you may take. So, I spend my entire life taking things, and of course, giving, but taking.

And so, when we met Mr. Ando we spent a lot of time eating fantastic sushi. Our office in Osaka was not far from his office during the Kansai Airport construction, so we were meeting quite often. And we very rarely talked about architecture. I don't think we spent a lot of time talking about architecture, first because he doesn't talk! (laughing) Second, because it's not very interesting. You talk about architecture as 'doing what you're doing'. You just make building and then that says enough. But having friends like Ando and others is essential because it gives you some feeling that you're not alone, you're living in the system of people. And the creativity is very important because it needs the work of being together. So it was

和意大利两个国家的文化传统都很深厚。但你也许会被那些令人叹为观止的遗产麻痹迷惑。你必须要谨慎。热爱传统决不意味着要去复制传统。复制只会带来感性的麻痹,是一件愚蠢的事情。你变得麻痹是因为你只关注美的东西。抢劫传统的惟一途径就是要变得足够好奇,有足够的创造力,正中求"变",从传统得到灵感。

我从传统得到的启发是关于材料、质感和建筑物的肌理、色彩、尺度,很多很多东西。日本是一个具有伟大传统的国家——我认为,实际上建筑设计、时装及任何事情应该成为传统,欧洲的情况也是类似,但并不总是如此。对于我来说,后现代主义❶是一种糟糕的经验!它很被动,并没有什么意思,这只是一种传

❶ 后现代主义:美国建筑史学家查尔斯·詹克斯最初使用的语言,是对20世纪后半期现代主义崩溃后出现的设计多元化倾向的总称。

圣·日内瓦图书馆内部
© K. Iwaki

STUDENT:
How do you think of tradition?
RENZO PIANO:
You know, love for tradition doesn't mean your reproduce tradition, of course. This is very important, because you then perform in the picturesque. I think Japan and Italy both have a very strong country culture where the tradition is very strong, important. But you may get paralyzed by that fantastic heritage. You have to be careful. Love for tradition never, never means that you reproduce that tradition. I mean literally. It's a bit stupid: you become paralyzed because this is not cared for, but just picturesque! The only way to rob tradition is by being curious enough, inventive enough, to 'change' from tradition, to be inspired by tradition.

My inspiration of tradition is about materials, it's about texture, it's about the grain of buildings, color, scale, so many things! Japan is a country with great tradition——I think architecture design, fashion, everything, should become actually tradition. The same thing in Europe, but it's not always like that. really great. Terrible and great!

统的置换。对后现代主义其实我是无计应对的。我喜欢一种流派,但它无路可走,实践也很局限。你们应该去努力翱翔,(后现代主义)只是一种文本上的复制,文本复制就像是一种照片复制,像一台照片复制机器。

此刻,你们正在谈论巴黎的蓬皮杜文化艺术中心,这是30年前——几乎30年前,很久以前——的一件具有非常强的挑衅性的作品。它从属于传统,巴黎的传统,追溯至上一个世纪,你就会发现铸造的钢构部件真实表达了巴黎的铸铁装饰❷传统。所以,某种程度上说,它属于巴黎,属于我们的历史本身,那时理查德·罗杰斯❸是我的合伙人,当时我33岁,他36岁,都属于年轻一代。我们是坏孩子,不受赞许的孩子,但是它确实是另类传统,因为这座建筑就像是巴黎中心的一艘船。这是我的看法,当然它有许多挑衅的东西,在巴黎中心区放置这样一个有趣的太空船本身就是一种挑衅。

❷ 铸铁装饰:例如,亨利·拉布鲁斯特(Henry Labrouste)设计的圣·日内瓦图书馆(1850年)、巴黎国家图书馆(1869年)的铸铁柱子以及拱券上的铸铁装饰。

❸ 理查德·罗杰斯(Richard Rogers,1933—)。英国建筑师,就学于伦敦的AA学校和美国的耶鲁大学。与诺曼·福斯特等人一起成立TEAM4,1970—1977年间和皮亚诺成为合伙人。代表作有"蓬皮杜文化艺术中心"、"Billingsgate Market"(伦敦,1988)等。

二十多岁的年轻人,希望你们有更多的求知欲

安藤——今天是一个难得的机会,应该让学生们多听

Honestly! For me, post-modernism was a terrible experience! Not very interesting because it was passive. It was just a transposition of the tradition. I have nothing against post-modernism. I love a good party, but I thought it was going nowhere, limiting the experience. You have to take the courage to fly. It was just a literal reproduction. And a literal reproduction is like a photo-copy, like a photo-copy machine.

Now, you were talking about Pompidou Center. Pompidou Center in Paris was about 30 years ago-almost 30 years ago, a long time ago-it was a very big provocation. It belongs to the tradition, Paris tradition. I mean the cast piece of steel is really in the cast-iron tradition of Paris if you go around to the previous century. So in some way this belongs to Paris, belongs to our history-myself, and Richard Rogers who was my partner then. We were young guys——I was 33 years old, Richard was 36, something like that. We were bad boys, very unappraised boys, first. But it was, in some way, traditional because Beaubourg (Pompidou center) is like a ship in the middle of Paris. And this is part of

一些皮亚诺先生的谈话。

我想今天在座的同学并不全是要成为建筑师的人，有人将来进入建筑业，也有许多人会进入其他公司。对我来说，二十多岁是一个非常重要的时代，正是因为有了二十多岁时的努力，才有了四十多岁时的成就，我认为二十多岁的积蓄会大大地左右一个人的人生。最后，我想请皮亚诺先生给我们一些提示，告诉我们二十多岁的时候最起码要做哪些事。

皮亚诺——好！如果你们还小于20岁，那么将进入你们的30岁，然后40岁再到50岁。我一直到60岁时才终于能够捍卫自己，所以这是一段漫长的旅程。我想，我能对你们说的是，从你们的20岁起就要像我刚刚讲的那样，你们应该捍卫属于你们自己的东西，而不是用所有的精力去捕捉信息。你们应该保持一定的时间沉静，这是我的第一点建议。所以，这就像在晚上睡觉前和早上晨起后喝杯水一样，你们每天至少要有20分钟时间一个人呆着，这也应是一个绝妙的想法。每个人都要试图去理解哪些是重要的东西，哪些是不重要的东西，这很要紧。另外就是我所说的"贪婪"，知识上的贪婪，好奇心是很基本的。看一看建筑界的人吧，如果没有建筑背后或底蕴中的东西去向上推动这座冰山的话，建筑就什么都不是。建筑如果是一种纯

my vision. So there's some tradition. There is Beaubourg, but there's a lot of provocation, and to put land with this funny spaceship in the middle of Paris, was part of the provocation.

TADAO ANDO:
Today, there are many students just beginning to study architecture. Would you give us some message for their twenties?

RENZO PIANO:
Well, if you are under 20, then you'll get in the 30s, then in the 40s, then in the 50s. I started to find my defense in the 60s, so it's a long journey. I think what I may say for you, since you're under 20, is what I said before: I think you should defend your privacy, not in terms of giving all energy to get information.

I think you have to stand sometimes in silence. This is my suggestion number one. So, it's like drinking fresh water in the night before going to the bed, or in the morning. Thank yourself for this beautiful idea, that you stay alone with yourself for at least 20 minutes a day. It's very good! Just to try to understand what is important and what

粹的学术，建筑就不存在了。这很可笑。

没有内容、没有冰山隐藏的部分，一切就很荒谬。无论是绘画、雕塑、音乐还是文学，如果没有内容就没有艺术。纯粹的学术一文不名，如此，建筑就更像是一种没有内容的建筑，这个内容就是生活！所以，你们必须具有旺盛的求知欲。我的意思是，当你们成长壮大时应该如此，当你们30、40乃至50岁时也是如此。等到你们60岁时就不需要了，因为你已经真正成为具有旺盛的求知欲的人了。

今天能够在这里以这种形式和大家座谈，我感到非常高兴。不过，安藤先生，我还有一个愿望，希望你们不要把我今天的讲话当作一份无用的讲义，到明天就扔掉！（笑）

安藤——（笑）非常感谢！

is less important. I think this is very important. Another thing is what I call "voracity," an intellectual voracity. I think curiosity is fundamental. Just look around at those who stay in architecture. Architecture is nothing without what is behind architecture or below architecture-to push the iceberg up. Architecture doesn't exist if it's pure academy. It's ridiculous.

All are ridiculous without content, without the rest of the iceberg. Painting, sculpture, music, literature, there's no art that means something without content. I mean, just pure academy, but that's bull shit. Architecture is even more like that-architecture without all that content. And this content is life! So, you have to be very curious. I mean, being curious when you're strengthening, means that you stay as curious when you're 30 and then when you're 40 and then when you're 50. Then, when you're 60 then you get impossible, because you really get very curious!

So, just please, Ando. Be sure that tomorrow you don't tell everybody the opposite of what I said! (laughing)

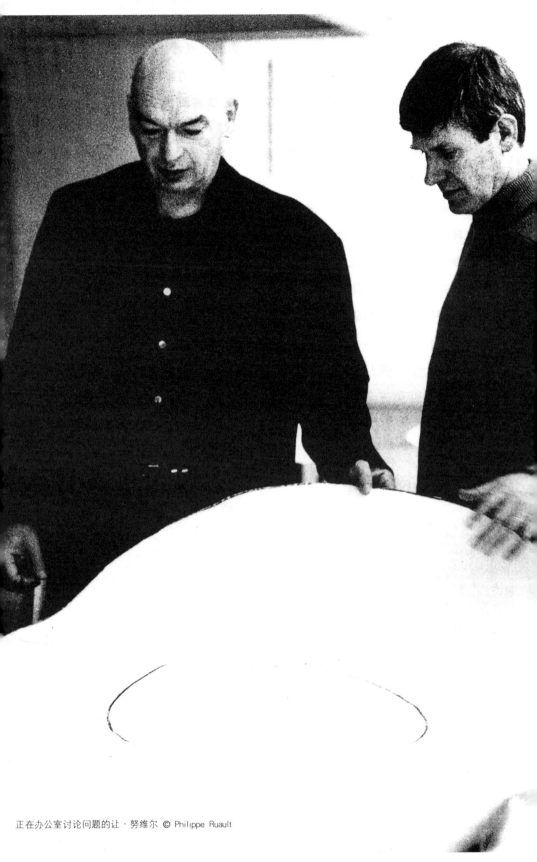

正在办公室讨论问题的让·努维尔 © Philippe Ruault

1998年6月18日（星期四）2:30pm — 4:00pm　主持：岸田省吾

让·努维尔
JEAN NOUVEL

　　努维尔1945年生于法国西南部的弗迈尔地区。1966年从波尔多转到巴黎的美术学院进行学习，次年起在克劳德·巴兰的手下工作，学习清水混凝土建筑设计。1970年成立事务所。从巴黎美术学院毕业后的数年中，一边进行住宅等设计，一边参加了难以尽数的设计竞赛。1971年以后，他通过为巴黎双年展设计舞台美术以及整体构成，建立了与艺术界的联系。

　　努维尔第一个为世人瞩目的作品是用铝合金外皮包裹起来的巴黎伯忠（Bezone）门诊医院。之后由于不断地进行剧场的设计工作，他获得了拜尔福特（Belfort）剧院的改造工程，在这个工程中，他在粗糙的石材中加入新的材料，强调外装饰材料的应用。

　　事务所成立11年后，他获得了"阿拉伯世界研究所"设计竞赛的胜利，这使他在法国建筑界崭露头角。该建筑将反叛的精神覆盖在优美的幕墙之下，得到广泛的认同。在之后的10年中，他得到许多机会尝试对建筑室内进行各种覆盖的设计方法。

　　通过操纵金属、木材、玻璃，并根据环境及文脉的需要考虑物质的三态，改变表皮的透明度，将建筑进行包裹。有时也会抓住外气进行利用。因此，他的建筑内部就像拥有生命一样，讴歌光和影。在里昂歌剧院(1983)、ONYX文化中心(圣·赫尔比亚／1989)、芬齐会议中心(图尔／1993)、卡梯尔基金会画廊(巴黎／1994)、拉芬耶百货公司(柏林／1996)、音乐会议中心(卢塞恩／1998)等设计中不断地进行着建筑师的尝试。

（三下晶子）

卡梯尔基金会画廊／1994 © T.Bojo

阿拉伯世界研究所／1987
© T.Bojo

我的第一位建筑老师，是一位17世纪的建筑师，他建造了我成长的家

❶ 阿拉伯世界研究所（巴黎，1987）：设计者是通过1981年举行的指名设计竞赛选定的。该项目是密特朗总统"巴黎计划"的第一环，向国内外展示巴黎建筑行政的剧烈变化。该建筑包括展示阿拉伯文明和伊斯兰文明的美术馆、画廊、图书室、音乐厅（拥有450个座位）、螺旋状书库等，向世界展示出法国的现代建筑风采，也使努维尔一举成名。该建筑的设计是将两个平行的体块沿东西向排列，围合建筑空间的是几种不同的幕墙：面向塞纳河的北侧是玻璃幕墙，面向广场的南侧是具有可调节功能的金属幕墙，面向中厅的是大理石墙面。

❷ 萨路拉（Sarlat）：位于法国西南部多尔多涅省（Dordogne）佩里格区（Périgueux），约一万人的小镇，旧街道中排列着中世纪和文艺复兴时期的住宅。附近的拉斯克洞穴（Lascaux）有公元前一万五千年前的壁画遗迹。

岸田——今天我们非常高兴请到法国建筑师让·努维尔先生来给我们做讲座。努维尔先生通过巴黎"阿拉伯世界研究所"❶的设计而一跃成为世界级的建筑师。这是20世纪80年代的一座著名建筑，从那以后他的活动大家都非常清楚，我想今天就没有必要再进行介绍了。

首先，关于您所接受的教育，我想提一些问题。我知道您的大学教育是在巴黎接受的，但直到高中教育为止，您是在什么地方、接受了什么样的教育呢？

努维尔——实际上，我不是巴黎人，我出生于法国西南部。从8岁开始，我的家搬到一个中世纪的小镇，名叫萨路拉❷，这个小镇的建筑在法国很有名。在那儿，我度过了几乎所有的少年时光。距离那儿20公里处有世界著名的史前岩洞，因此欧洲人都知道萨路拉这个小镇。我经常说我是史前人，可能那个时候的什么东西，一直残留在了这里（指着眉间）。我一直有一种感觉，好像自己生活在原始时代，因此，唯一憧憬的就是未来。

SYOGO KISHIDA:
J'aimerais d'abord vous poser quelques questions à propos de l'éducation scolaire que vous avez reçue. Quel chemin avez-vous suivi jusqu'à l'ensignement secondaire?

JEAN NOUVEL:
Effectivement, je ne suis pas Parisien. Je suis quelqu'un qui est originaire du Sud-Ouest de la France. Et j'ai passé presque toute mon enfance, à partir de l'âge de 8 ans, dans une petite ville d'architecture, une petite ville du Moyen-Age, qui est très célèbre en France et en Europe, qui s'appelle Sarlat. Et qui est à 20 km d'un lieu mondialement connu, puisqu'il s'agit des grottes de Lascaux, les grottes préhistoriques. Et je dis souvent que je suis un homme préhistorique. D'ailleurs, il m'en reste quelque chose là… Et je dis souvent que la seule nostalgie que j'ai, c'est celle du futur, parce que j'ai vraiment l'impression de vivre une époque préhistorique.

Quand je vois déjà tout ce qu'on a pu inventer en deux siècles, je me dis qu'il faudrait mieux vivre dans 20, dans 50 ou dans 100 siècles. Je suis quelqu'un de relativement optimiste, et je pense que notre vie est meilleure que celle

从19世纪至20世纪这两个世纪人类的各种各样发明来看,我们可以相信,在将来的20个世纪、50个世纪甚至100个世纪以后,人们的生活一定会更好。我是一个乐观的人,我认为虽然现在世界上还有许多不尽满意的事,但我们的生活已经比若干个世纪以前好很多了。

但是,不管怎么说,我还是原始时代的人,而且,还有中世纪的影响。为什么会这样呢?这与我居住的城市有关。萨路拉这个城市曾经进行过整体的修复,它完全再现了中世纪的街区形象。即使在法国,它也是一个非常特别的城市。

我8岁时来到萨路拉,住进了一座17世纪的贵族宅邸。这幢老房子里留存着那个时代的一些东西。楼梯是用石头砌成的,上面环绕着铸铁扶手,梯面的一部分已经被磨下去了。楼梯顶部的木质吊顶让我非常着迷,上面绘有壁画。虽然画面颜料已经发黑,一切都隐没在灰暗之中,但仍然能隐约识别出天使以及半裸的女人,浮在上面。这是非常有刺激性的,它就像电影中的下意识作用一样,无形中在我的心中刻下了印记。也许正是因为这个经历,使我形成了这样一种思考方式,就是经常去深入观察并探索事物的深层含义。

这栋建筑里还有很多彩绘玻璃窗,由12厘米左右

qu'avaient nos semblables il y a quelques siècles. Même si un long chemin reste à faire surtout à l'échelle planétaire.
Bon, toujours est-il que je suis un homme préhistorique. Et moyenâgeux, puisque Sarlat est une ville du Moyen-Age, qui est une ville qui a fait l'objet d'une attention particulière en France puisque c'est une ville qu'on a entièrement restaurée, pas uniquement chaque monument, mais pour créer un ensemble architectural qui redonne les caractéristiques de ce qu'était une ville du Moyen-Age.
Et donc, dans cette ville, j'ai débarqué a l'âge de 8 ans, et mes parents sont allés dans un hôtel, ça s'appelle comme ça, une vieille maison noble du 17e siècle. Et dans cette vieille maison du 17e siècle, il y avait des choses extraordinaires: il y avait une grande cage d'escalier en pierre avec toutes les marches usées, avec des rampes en fer forgé avec des gros bouts de fer comme ça très larges et très corrodés. Et il y avait une chose qui me fascinait totalement, c'est qu'il y avait en haut de cette cage d'escalier un plafond peint en bois complètement caramélisé à cause des vernis… et tout ça était très mal éclairé, mais on pouvait deviner de façon quasi-subliminale

见方的小块拼合，并用铅丝固定而成。在夜间，窗子的内侧木百叶可以关闭。房间的地板由多种木材拼成各种图案。

如果要回答刚才的问题，我想我的第一位建筑老师应该是这位17世纪的建筑师，他建造了这栋宅邸。因此，可以说在我学习建筑之前，就已经从这个家中受到了不少建筑的熏陶。

与德维耶老师的相遇，改变了我的人生

少年时期

努维尔——如果要谈我所接受的具体的教育的话，我想必须首先讲一下，我的父母都是教师。父亲教英语，同时也是学校的监督员。做他们的儿子可不是一件舒服的事情。同学们总是嘲笑我，我也必须比别的同学更加努力一些，才能不给父母脸上抹黑。更糟糕的是我母亲站在讲台上提问的时候，如果我回答不出来，就会听到她说："明天课堂上还会提问的。"其他的老师也常常故意提问我，我经常要比其他同学接受更多的提问。

因为父母是教师，他们当然希望我将来能成为工程师或者教师。在他们看来，重要的是学习数学、法

dedans des anges,une allégorie d'une femme a moitié nue.... tout ça m'intriguait beaucoup et je pense que ça m'a peut-être conditionné,dans le sens que j'ai toujours essayé de voir en profondeur dans la matière.

Il y avait aussi des vitraux,avec des petits bouts de verre qui devaient être grands comme ça,12 cm,et qui étaient reliés par du plomb. Avec des petits volets de bois intérieurs qui permettaient de tout fermer. Il y avait aussi des planchers avec des marqueteries de bois,de différents bois. Tout ça pour dire que ma première éducation finalement,mon premier professeur,a peut-être été un architecte du 17e siècle. En tous les cas, avant de savoir que je deviendrais architecte, j'ai été profondément questionné par cette maison. Alors,en ce qui concerne mon éducation à proprement parler,il faut déjà savoir que j'étais le fils d'enseignant. J'étais le fils de l'inspecteur des écoles et d'un professeur d'anglais. Mais être fils de l'inspecteur dans une école,c'est jamais très confortable. Les copains sont toujours en train de vous charrier. Donc j'ai toujours été obligé d'en faire un peu plus pour me dédouaner, pour montrer que je n'étais pas là uniquement pour répercuter toutes nos bêtises a nos parents.

语等主课，还有历史、地理等科目，而音乐、绘画之类的科目在他们看来根本就不重要，因此，小时候我根本就没有接受过艺术类的教育。

在大学考试的两年前，也就是高中一年级的时候，我遇到了一个美术老师，他改变了我的一生。他就是德维耶老师。在法语中，德维耶就有"选择今后的方向"的意思。（笑）

为什么说这个老师改变了我的人生呢？我最初画的一幅画，我自己认为非常糟糕，可他看后却表扬了我，认为我很可能有这方面的天分。从此，我渐渐地开始对绘画感兴趣，并钻了进去。德维耶老师还邀请我去参加他个人办的静物画和人物画的培训班。在那里我有了自己的画板，碳条，红粉笔，毛笔，刮刀。但是，每隔两三次会有一次裸体写生，他们总以我太年轻为由把我赶出来，这让我很不高兴。那时候，我总是准确地计算好时间到画室，为的是能够看一眼模特的进出。（笑）

从画家到建筑师

努维尔——从这以后，我就想成为一个画家，并想进

Mais ça a fait que je n'ai jamais eu le choix:j'ai dû travailler. Puisqu'à chaque fois que je ne savais pas une leçon,ma mère me disait: "On verra ça demain,en cours."Et tous les professeurs mettaient un malin plaisir à me questionner,à demander plus de choses à moi qu'aux autres.
Mais en tant que fils d'enseignant,mes parents voulaient que je devienne ou ingénieur ou professeur,ou quelque chose comme ça. Et les choses importantes étaient les mathématiques, le français,les matières principales,l'histoire,la géographie... Par contre,je n'avais absolument eu aucune éducation axée sur les disciplines artistiques. Musique et dessin,ça n'existait pratiquement pas. C'étaient des matières secondaires.
Et quand je suis arrivé en seconde,deux années avant le BAC,je suis tombé sur un professeur de dessin qui a complètement dévié ma vie. D'ailleurs,il s'appelait Desviers,c'était son nom.
Alors,pourquoi il m'a dévié,cet homme? Parce qu'il a eu l'intelligence de me faire croire que mes premiers dessins-c'étaient des études académiques en classe-avaient une certaine

① 美术学院：在法国，高中毕业后，如果想继续接受高等教育，可以选择进入综合大学或者高等学院。如果要进入综合大学，就必须在高中的最后一年参加考试取得入学资格。要进入高等学院就必须在高中毕业后先进入预备学校学习1～2年，然后通过书面审查，再接受各学院的入学考试。高等学院与综合大学的教育方针完全不同，是一个培养技术人才的教育机构，有师范学院、行政学院、工学院、音乐学院、美术学院等。美术学院统合了建筑、绘画、雕刻等分校，从1819年开始进行教育。但1968年以后，建筑学院又独立出来，另成一所学校。

入美术学院①的绘画专业进行学习。给父母讲了我的想法以后，他们严厉地告诉我："如果你要学绘画，那我们是绝对不会给你学费的。靠画画这种职业，是绝对不可能生活下去的。再认真考虑一下，拿出一个像样一点的方案来。"

从那以后我也改变了战术，说："知道了，我会认真地学习数学和物理。但同时，我也想学一些建筑。"（笑）当时我在心中暗暗盘算，先设法进入美术学院学建筑，过一两年后再转向美术。这样，进入大学的两个月后，我以建筑专业的学习课程太重，不可能同时学习数学、物理为借口说服了父母，顺利地得到了他们的许可，专心地学习建筑。那时候的法国，有美术学院和大学两种体系，为了学习建筑，我进入了美术学院，放弃了大学。我在美术学院学习的时候是一边学习建筑一边工作，两件事都做。

刚开始的时候，在波尔多美术学院。那是一个古老的学校，有点过时了，到那里去主要是为了获得自由。对我来说，自由就是离开父母身边，开始大学生活，也可以享受夜生活了。我的零花钱都在玩扑克的时候输掉了，而且每天还玩橄榄球。波尔多美术学院有老生欺负新生的恶习，这是大家都知道的事。那时学校最重视的就是橄榄球队，于是我就对那些欺负新

beauté. Alors que c'étaient des dessins totalement minables. Mais petit à petit, je me suis passionné, et j'ai commencé à dessiner, à dessiner de plus en plus. Et il m'a invité à ce moment-là à des séances privées où ils dessinaient ou peignaient des natures mortes ou des modèles. Alors, j'avais mon chevalet, mes fusains, mes sanguines, mes pinceaux, mes couteaux… et j'étais très vexé parce que, sur une séance sur deux ou sur trois, j'étais évincés parce que c'étaient des modèles nus, et je n'avais pas le droit de voir des femmes nues à cette époquelà. Je m'arrangeais quand même pour arriver quelquefois au bon moment, pour les voir partir ou arriver.

Et à la suite de ça donc, j'ai voulu devenir peintre. Et entrer aux Beaux-Arts en section peinture. Et là je suis tombé sur un tout petit problème, c'est que mes parents m'ont dit: "Pas question, nous, on paie pas des études aux Beaux-Arts parce que tu vas passer ta vie à tirer le diable par la queue, ça veut dire ne pas gagner d'argent. Donc, propose-nous quelque chose de sérieux."Et dans tous les cas de figure, je devais faire des mathématiques, en études supérieures. Et là, j'ai élaboré ma stratégie, j'ai dit: "Certes, je

生的老生们采取这样的态度：如果你们做得太过分的话，我就不参加橄榄球队。终于，到了一个季度快要结束的时候，也就是五月，我受到了报复。一次是被剃光头，也就是和我现在这个样子一样，另一次是用红药水从头涂到脚尖，全身都是红的，怎么样，想像得出来吗？（笑）

在克劳德·巴兰和保罗·维利利奥的事务所学习

努维尔——后来，我去了巴黎。由于在波尔多根本没怎么学习，所以只能算是一般的水平吧。即使这样，我仍然是同届学生中少数几个能够参加巴黎美术学院入学考试的学生之一。因为我不怎么到学校来上课，这在校内还成了一个很大的话题。

这样我考取了巴黎美术学院，进入当时最年轻的老师的工作室❷。一年后，因为我想在20岁的时候结婚，所以就决定一边学习一边在事务所工作挣钱。我在美术学院入学考试中考了第一名，这个名次起了作用。要知道，入学考试是非常难的，在1200～1500名考生中只录取10%。由于我考了第一名，所以立即就进入了建筑师克劳德·巴兰❸和保

❷ 工作室：当初在美术学院创立的建筑学专业有这样一个传统：学校只进行课堂教育，建筑实习是在学校以外的建筑师工作室进行。这些工作室大部分都采用一种类似私塾的体制。1963年美术学院大改革之后，在以往的工作室的基础之上，校内又增加了三个由主讲教授负责的公立工作室。不过，1967年工作室制度解体。

❸ 克劳德·巴兰（Claude Parent, 1923—）：生于讷伊（Neuilly），其父为建筑工学者，就学于巴黎美术学院。曾与勒·柯布西埃一起工作。作品有"奴依依自家住宅"（1963）、"伊朗学生会馆"（巴黎，1968）等。作品中常使用具有雕塑感的粗混凝土造型，给人强烈的印象。著作有"在倾斜中生存——城市规划中的冒险"（Vivre à l'Oblique, L'Aventure Neuilly,1970）、"关于建筑的五个考察"（Cinq Réflexions sur l'Architecture, Nevers, Maison de la Culture,1972）等。

vais faire mathématiques et générales-physique, puisque je suis obligé,mais je vais faire aussi architecture. "Et avec la ferme intention,au bout d'un an ou deux,comme j'étais aux Beaux-Arts, de revenir en peinture. Donc,deux mois après mon entrée à l'université,j'ai expliqué a mes parents que je ne pouvais pas faire à la fois mathématiques,générales-physique et architecture,parce qu'architecture,c'était trop de travail. Donc,j'ai obtenu l'autorisation de me consacrer à l'architecture. Puis,j'ai décidé très vite de gagner ma vie. Pour des raisons diverses et variées. Et à partir de là,donc,j'ai décidé de faire architecture,et j'ai commencé à travailler, en parallèle de mes études,à faire la place comme on dit... Alors là,j'ai commencé mes études à Bordeaux,mais Bordeaux était une école un peu ancienne,un peu démodée.

Alors,le départ à Bordeaux,c'était d'abord la liberté. La liberté,ça voulait dire être à l'université sans les parents,et j'ai passé mon temps à vivre la nuit déjà... à commencer à vivre la nuit. A dépenser le peu d'argent que j'avais au poker. A jouer au rugby,ce qui avait un énorme avantage parce que l'école des Beaux-Arts était une école très archaïque,je vous l'ai

❶ 保罗·维利利奥 (Paul Virilio, 1932—)，1964年与克劳德·巴兰相识，一起提出"倾斜的功能"学说，并在建筑中开始实践。著作有"削减的美学"(Estbétique de la Disparition, Paris, Balland, 1980)，"战争与电影"(Guerre et Cinéma, Paris, Éditions de l'Etoile, 1984)等。

罗·维利利奥❶的事务所工作。我相信，对我来说，真正的建筑教育应该是从这一天开始的。

克劳德·巴兰一开始就非常喜欢我。那时，我才20岁，事务所大概有20个人。一年后，对建筑还什么都不知道的我却被任命为一个80户公寓的建筑项目的负责人。由于所里还有资格更老的人在，所以这件事让全体员工都非常吃惊。

在工地上，施工人员和工程师详细地告诉我混凝土的浇筑方法、配筋、门窗、换气等各种事情。我当时就像一个什么都不知道的孩子一样，他们非常耐心地教给我许多事情。在法国以前有这样的一种说法，把孩子扔到水中他就能学会游泳，我觉得我就是受到了这样一种教育。

和保罗·维利利奥
闲谈度过下午的时光

努维尔——第二个幸运之处，就是遇到了保罗·维利利奥。克劳德·巴兰这位建筑师不是一位常呆在事务所里的人，而保罗·维利利奥则每天下午都在事务所。在这样的下午，保罗·维利利奥会和我谈论和建筑没有直接关系的无边无际的话题，例如军事方面的事情、圣经最

dit, et très caractérisée par des bizutages très forts. Et une des choses les plus importantes a l'école était l'équipe de rugby. Et j'avais fait comprendre aux anciens qui me bizutaient que s'ils étaient trop durs avec moi, je viendrais moins au rugby, et je m'intéresserais moins au rugby. Mais dès que la saison de rugby a été terminée, en mai, ils se sont vengés. Alors, j'ai eu droit à une préfiguration de ce que je suis devenu. La grande spécialité, c'était la tondeuse. J'ai été tondu dans tous les sens, partout, c'était absolument terrible. Les grandes spécialités aussi, c'était le passage en broche, c'est-à-dire qu'on vous passait du mercurochrome du haut en bas, vous étiez rouge comme ça. Vous imaginez le spectacle.

Toujours est-il donc, après je suis parti à Paris. A Bordeaux, ça s'était passé, je dirais, de façon assez moyenne, parce que j'avais peu travaillé, mais j'avais quand même été un des rares élèves de ma promotion à être présenté au concours d'entrée des Beaux-Arts. Ce qui avait provoqué des réactions, parce que j'étais pas là souvent.
Et je suis allé à Paris dans l'atelier où le patron était le plus jeune de France. Et au bout d'un an, parce que j'avais décidé à l'époque de me

后一章默示录中出现的末日思想,也就是人类末日的话题等,这些闲聊伴随我们度过了许多下午的时光。

我不知道你们是否知道保罗·维利利奥这个人,他在法国是非常有名的。他写的关于战略思想的书甚至在民间都成为话题。特别是那本与战争有关的《掩体考古学》❷一书非常有名,书中写了战争时必需的要塞、军火库等形态与建筑的关系。在书中他多次写到,军队拥有强大的财力,在技术上也总是处在最先端的地位,因此,在他们建造的建筑形式中,蕴含着下一个时代的技术要素。

❷《掩体考古学》:Bunker Archéologie, Paul Virilio, Paris, Les Éditions du DemiCercle, 1991.

他实际上关心的事情非常多,还写过关于速度的书❸。也就是说,他考虑了信息流通的加速给这个世界的结构带来的变化。

❸关于速度的书:《速度与政治——从地政学到时政学》Vitesse et Politique, Paul Virilio, Galilée, 1977.

下午呆在事务所的保罗·维利利奥先生,对我来说,是在建筑老师之外的另一种意义上的一个非常重要的老师。他总是对我讲他自己关心的各种事情。有时,也会因为没有做自己的本职工作——设计,只是听保罗·维利利奥讲话,受到克劳德·巴兰先生严厉的训斥。(笑)

那个时候,克劳德·巴兰和保罗·维利利奥设计了一个像要塞一样的教堂。这个名叫"圣·拜尔纳提德"❹的教堂是他们的作品中非常有名的一个,我想安

❹圣·拜尔纳提德教堂(Ste.Bernadette,1966):不规则形状,几乎没有开口部分,外观就像一个大型的混凝土块,圣堂设在二层,形体倾斜并向外悬挑,圣堂的地面、座位排列、平面、墙面、吊顶、开口部等都大量采用了斜向的要素。维利利奥称这个作品是他将倾斜的理论用于实践的第一个作品。

marier à 20 ans,de me marier pour quelques mois,c'était juste... J'avais décidé de beaucoup travailler. Et là,il s'est passé une chose qui m'a beaucoup aidé,c'est que j'ai été reçu numéro un au concours national d'entrée à l'école des Beaux-Arts. Il faut savoir qu'il y avait à peu près 1200 à1500 candidats chaque année,et 10% qui étaient pris. Pourquoi ça m'a aidé? Ça m'a aidé parce que ça m'a permis d'entrer après, très rapidement chez Claude Parent et Paul Virilio. Alors,je dirais que c'est là que ma vraie éducation,puisque vous vous intéressez à ça,ma vraie éducation a commencé ce jour-là je crois.

D'abord,parce que Claude Parent m'a tout de suite pris en affection. J'avais 20 ans,j'étais dans une agence où il y avait une vingtaine de personnes,et au bout d'un an,alors que je ne savais rien,il m'a nommé chef de projet sur un projet de 80 logements à Neuilly. A la surprise absolue de toutes les personnes de l'agence,dont certaines y travaillaient depuis longtemps.
Et les entrepreneurs et les bureaux d'étude sur le chantier m'ont pris par la main le soir pour m'expliquer ce que c'était que le béton,les ferraillages,les menuiseries,les ventilations etc. J'étais tellement un bébé que ça provoquait de

53

❶ 倾斜的功能：该学说认为倾斜的空间可以诱发建筑本身内在的运动。"建筑的原理"一书中有宣言全文。

藤先生也会知道。当时他们就"倾斜的功能"❶这一建筑原理展开讨论。他们构思了这样一种形态：墙面不是垂直的，也不是正交，所有的空间都是以倾斜的方式联系在一起。他们用这种方法建造了许多作品。

当时他们两个人建造的建筑都是清水混凝土建筑，因此，我年轻的时候也是生活在清水混凝土的世界里。当时使用的是木模板，为了在混凝土的表面表现出各种纹理，他们进行了各种各样的尝试。例如他们曾经尝试用磨砂的方法磨去模板的柔软部分让木纹更加明显，也研究怎样连接模板才能够不留下痕迹等等，十分认真刻苦。

改变我一生的 1968 年的五月革命

❷ 1968年，1968年5月—6月，一场涉及政治、经济、文化领域的运动从法国波及到欧洲，称为"五月革命"。

❸ 兰博基尼是创始于1962年意大利的著名跑车公司。
——译者注

❹ 捷豹，也译成积架，英国著名的汽车制造公司，旗下还拥有著名的跑车、赛车系列。
——译者注

努维尔——随后，对我的一生来说，一个非常非常重要的时刻来临了。

1966年入学后，很快就迎来了1968年。"68年"❷对巴兰和维利利奥的事务所来说也是一个非常重大的事件。当时，他们住在巴黎最高级的住宅区中，巴兰开着兰博基尼（Lanborghini）❸，维利利奥开着捷豹车（Jaguar）❹。维利利奥很快扔掉了捷豹车，全身心投入

l'affection. C'était en fait la vieille technique de prendre le bébé,de le foutre à l'eau et de voir s'il sait nager.
Alors, il y a eu une deuxième chose formidable, c'est que Claude Parent était pas souvent à l'agence,pas trop souvent,et que Paul Viriliot, lui,y était tous les après-midi. Et donc,Paul Viriliot passait son temps les après-midi à me raconter toutes ces histoires complètement folles sur toutes ses déviations,sur la chose militaire, toutes ses visions apocalyptiques sur le malheur qui nous attendait etc.
Alors,je ne sais très bien si vous savez qui est Paul Viriliot,qui est très célèbre en France. Et Paul Viriliot a écrit après des livres qui sont maintenant très connus où il a mis en avant des idées stratégiques,je dirais. Il a fait en particulier "Bunker archéologie",c'est lui qui s'est intéressé le plus près et le plus tôt à toutes les formes d'architecture militaire. L'une de ses idées fortes,en particulier est le fait qu'il faut s'intéresser à la chose militaire parce qu'elle est toujours prémonitoire. Comme les militaires ont toujours beaucoup d'argent et toutes les technologies à leur disposition,y compris celles qui restent secrètes très longtemps parce qu'elles ont

un intérêt militaire,on voit souvent dans les objets militaires réalisés la préfiguration du futur dans d'autres domaines. Oh!il s'est beaucoup intéressé aussi,il a écrit un livre sur la vitesse. Sur le fait que la vitesse a changé complètement notre monde à travers toutes les données les plus immatérielles sur les réseaux, sur la vitesse de l'information,sur la ville où la nature de l'espace est beaucoup moins importante que toutes ses connexions etc.

Toujours est-il que Paul Viriliot,ce personnage absolument incroyable,était mon professeur toutes les après-midi. A tel point qu'il se faisait engueuler par Claude Parent parce que j'étais là pour travailler,et qu'en fait quand Paul Viriliot était là,je ne foutais rien. Donc,c'était l'époque où ils construisait en particulier l'église sous forme de bunker de Ste. -Bernadette de Nevers,je pense que Tadao connaît ça. Et où ils développaient toutes les thèses d'architectureprincipe sur la fonction oblique,la fonction oblique étant l'élimination de la verticalité et l'organisation spatiale de la cité, mais aussi des immeubles à partir des plans inclinés parcourables et vivables.

Alors tous les deux étaient les apôtres en France

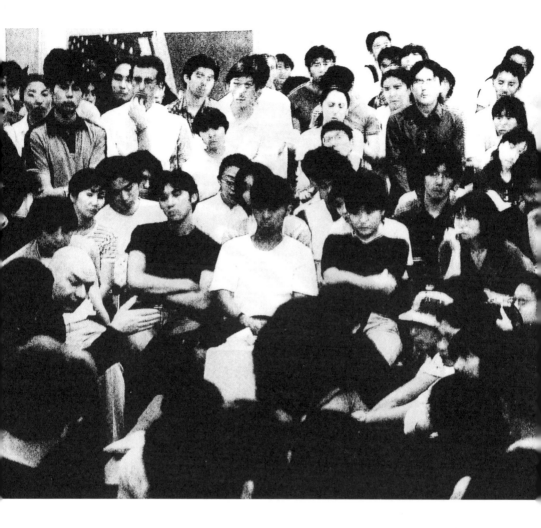

du béton brut. Donc j'ai commencé ma vie dans le béton brut. A savoir comment on coffrait ce béton, avec des planches qui étaient sablées pour montrer bien les veines du bois, c'était d'abord un travail de charpente, pour bien montrer toutes les empreintes…
Et puis il s'est passé une chose incroyable à l'échelle de ma vie.
J'ai commencé en 66, et très vite, ça a été 1968.
Et 68 a été un gros événement à l'agence Parent et Virilior parce que Parent roulait en Lanborghini et Paul Viriliot en Jaguar, tout ça, ça se passait à Neuilly, autrement dit dans le quartier le plus luxueux de Paris. Paul Viriliot a laissé sa Jaguar et est entré dans Mai 68 à fond la caisse, et c'est lui en particulier qui a organisé la prise de l'Odéon, qui a été un grand événement de 68, du Théâtre de l'Odéon. Mais comme Viriliot était un général rentré, tout son sens stratégique s'est développé là, il y avait des articles dans la presse, je me souviens "Viriliot, le diable Kohn Bendit". Kohn Bendit, c'était le leader de Mai 68.
Et moi j'ai vécu Mai 68 comme la réalisation de quelque chose que j'attendais sous cette forme ou sous une autre. Déjà, je vivais mon expérience

Parent & Viriliot comme un rejet ou comme une contestation directe de l'Ecole des Beaux-Arts. A l'Ecole des Beaux-Arts, je me suis rendu compte très vite qu'on nous apprenait que des recettes. Et des recettes qui pour l'essentiel était des recettes d'ordre purement graphique. Alors, le grand sport de l'époque, c'était de monter en loge, quand on avait des exercices importants, on restait 6 heures, 8 heures, 12 heures, 48 heures des fois bloqué, enfermé pour faire tout seul son dessin. Le grand exercice, l'exercice le plus haut, c'était le Prix de Rome, là on pouvait passer, quand on était dans les derniers logistes, on passait je crois 3 semaines en loge pour faire le grand projet du Prix de Rome. A la fin, c'est 3 semaines pour les 6 derniers.

Mais je me suis rendu compte que ce qu'on nous apprenait, parce qu'il fallait voir ce que c'est que les montées en loge, les gars avaient des tas de poncifs, c'est-à-dire qu'ils dessinaient sur du calque, avec du crayon, des choses qu'ils n'avaient plus qu'à poncer après. Il y avait des nuages, des rendus de nuit, des arbres, des dallages, toutes sortes d'éléments qui permettaient de composer une image extraordinaire. Tous les éléments pour les titres aussi, les pochoirs...

到1968年的五月革命中去了。"68年"事件中的一件大事就是奥迪恩（Odeon）的占领，而进行策划的人就是维利利奥。维利利奥曾经当过军人，具有战略上的本领。我记得报刊上把他和"68年"的领袖孔·班迪（Kohn Bendit）一起做了这样一个报道《维利利奥、恶魔孔·班迪》。

当时我并没有非常清晰地意识到"68"年的真正意义，只是预感到自己所希望的事情可能会实现。本来，在巴兰和维利利奥的手下进行实际工作，就是和革命前的美术学院的教育体系完全相反的生活方式，因此，对我来说，在某种意义上，只是在时间上碰巧赶上了"68年"。

进入巴黎美术学院以后，我立即就意识到这里只会教给我一些技法上的东西。这个学校每年都在重复同样的东西，也就是对图纸进行单纯地描绘，用这样的方法进行建筑教育。

当时，教学中有一个固定的一个方式，就是课题一出来，就在一个封闭的环境中，尽可能长时间地做这个课题，短则6小时、8小时，长则12小时，有时甚至48小时一个人关起来画图。那个时候美术学院的奖项中最有名的罗马大奖❶，就是一个长时间的严酷的课题，如果进入最终选考，就要花上三周的时间进行设计作业。

❶罗马大奖：该奖项的设立是为了选拔优秀的学生，给他们留学罗马的机会，从绘画、雕刻、建筑三个专业中各选出一名学生，该奖在旧美术学院工作室制度下竞争非常激烈，1968年随着旧美术学院的解体而取消。

Une bonne partie de l'énergie,c'était la façon de composer la feuille,le panneau qui était rendu. D'abord,parce qu'il y avait déjà la perversion de l'anonymat,et que de la façon dont on présentait un panneau,on savait vous reconnaître. On savait reconnaître de quel atelier vous veniez,s'il y avait un noir en bas,s'il y avait le bleu qui traverse,s'il y avait autre chose... C'était une marque de fabrique.

Et tout cela me paraissait strictement opposé au fond des choses. C'est-à-dire qu'on était là pour proposer un projet,et pas d'abord sa représentation.

Et parce que j'avais été reçu nol et que j'avais fait la preuve que je savais très bien dessiner, aussi bien les plâtres que toutes les représentations architecturales avec les jus et tout ça,tout le monde savait que je savais faire ça.

A partir de là,je me suis payé deux provocations. La première,c'était une étude analytique sur les tours. On nous obligeait à rendre compte d'une typologie de tour,qui correspondait aux choses les plus "corporate"du style international. Les tours. Une étude analytique,c'est-à-dire qu'on devait prendre un exemple et le développer. Et

那个时候所做的事，只是完成用模板来画图这样一种作业。制图模板多种多样，其中也有夜里的景色，和云彩、植物等，谁能够准确地运用它们在图纸上画出优美的图形谁就能成为第一名。例如，在绘制标题的时候，要使用多种模板，运用多种技法，能够运用这些技法画出优美的构图的作业就会得到很高的评价。在图面构成中怎样使用技巧，来形成优美的构图，这种非本质性的问题被放在了非常重要的位置。虽然当时采用的是工作室制度，而且也分成几个工作室进行作业，但是一看作品，就知道是哪个工作室的东西，他们采用的教育方法都是只让学生学习图面技巧。黑色在下面的是这个工作室，中间有一块蓝色色块的是那个工作室，就像制作商标、图标一样。

向美术学院式教育方式挑战的两个方案

努维尔——我认为，让学生自己思考问题、自己进行方案设计才是本质性的教育。巴黎美术学院的做法充其量也只不过是学习了表现方法。我是作为第一名入学的，毫不客气地说，这些东西我已经会了。我已经完全明白如何进行图面表现，也就是说仅仅拥有一些

à partir d'une étude du texte,j'ai proposé de montrer comme étude analytique au lieu de ces tours-là,le tour d'attaque romaine sur roues tirée par des chevaux,ces tours qui montaient pour aller à l'assaut des fortifications. Entièrement en bois. Tout ça tiré par des chevaux et dessiné "à la Léonard de Vinci". Ce qui était une façon directe de me foutre de leur gueule.
Et effectivement,mon projet a été refusé. C'est le genre de plaisanterie qui coûtait 3 mois d'étude à chaque fois,puisque le projet était refusé.
Et la deuxième provocation,qui celle-là était peut-être plus signifiante,c'est-à-dire plus positive,parce que l'autre était purement négative. C'est qu'au lieu de rendre les 3 panneaux comme demandé,format 2 m par 1 m, pour un projet long,pour un grand projet,j'ai rendu un projet dactylographié. Basé sur une étude sociologique,très très étayé, une véritable thèse,avec un de mes amis qui faisait sociologie dans une école à côté,et qui à la suite de ce scandale en 67 et après 68,est devenu professeur à l'Ecole des Beaux-Arts. C'est mon ami Routon. Et ce projet,c'était la description précise de ce qu'était le projet,c'est-à-dire que par les mots,

绘画技巧是一件没有意义的事情。

因此，我进行了两次大胆的挑战。一次是在对当今的超高层建筑进行分析的课题中，另一次则是在一个提出综合性方案的三个月的课题的时候。这时只有我一个人，尝试了逆向分析的方法，我没有按照课题要求进行分析，而是对文献进行了彻底的阅读，提出了一个罗马时代的要塞那样的塔的方案，用木头制成，基座下面有轮子，用马来拉着走。当然，方案也是用达·芬奇风格来描绘的。这明显激起了教师们的愤怒，自然遭到了拒绝。学校是不容你这样开玩笑的，这三个月的工作等于白做了。

第一次我直接运用了这种消极的方法，但是第二次我想我应该用一些稍微积极的方法。在一个要求做成三个2米×1米图版的课题中，我采用了用文章的方式提交了这次方案。在文章中我进行了社会学的分析，用论文的形式进行了组合。其间我得到了学习社会学的朋友鲁顿的帮助，完成了这次作业。他在1967年的事件以及"68年"之后，成为了巴黎美术学院的教授。

具体题目是"为了儿童的图书馆"。请社会学朋友帮助分析，主要是想调查不同地区的学校，了解孩子们头脑中对图书馆的印象。在调查中，我们委托老师

c'était un vrai projet,c'était pas autre chose. C'était une bibliothèque pour des enfants.
Ce que j'ai demandé comme analyse à mon ami sociologue,c'est de questionner à travers différentes classes de différentes régions quelles images avaient dans la tête des enfants de ces régions. Et donc les instituteurs ont fait faire aux enfants des dessins de leur bibliothèque,de ce qu'ils imaginaient. Et on s'est rendu compte que les enfants reproduisaient directement ce qu'ils voyaient autour d'eux. C'est-à-dire,en Bretagne où les maisons sont comme ça avec un toit en ardoise et des murs blancs,la plupart des enfants faisaient leur bibliothèque comme ça. Dans le Sud,comme c'est beaucoup plus plat etc. ils les faisaient tous pareils. A Paris,c'étaient des petits cubes dans la banlieue etc. Tout ça pour montrer que notre imagination est conditionnée par notre information.
Et à partir de là,j'ai développé un projet qui montrait des variétés d'espace de nature à ouvrir l'imaginaire des enfants. Donc c'était un peu basé sur les Cent Fleurs,c'est une sorte de fleur et chaque espace avait ses caractéristiques.
Mais ce qui est important,c'est d'avoir proposé quelque chose de nature conceptuelle face à des

让孩子们画出他们想像中的图书馆。从结果中，我看出他们都是把自己身边所看到的东西直接拿来进行描绘。在布鲁塔尼地区，有许多双坡瓦屋顶和白墙构成的住宅，这一地区的大部分孩子就画了双坡屋顶的图书馆；平屋顶建筑较多的南部地区的孩子就画了平顶建筑；而巴黎的孩子则画出了现代建筑那样的东西。我得出了这样一个分析结论：我们的想像是被信息所左右的。

我的方案就是在这个分析结论的基础上做出的，我设计了许多变化的空间，希望能够增加孩子们的想像力。我考虑用不同类型的空间组合形成一个整体的建筑。实际上，当时我的目的是希望针对那些形式主义的建筑师们，拿出一些概念性的东西，来对抗他们。

被退回来的毕业设计

努维尔——当时的一个倾向是，做这样的方案也是可以的，但并不是学校应该鼓励的一个方向。但是，"68年"改变了这一切，在那一年我也建成了自己的第一个作品❶，是一个小住宅。这座建筑明显受到巴兰和维利利奥的强烈影响，同时也有詹姆斯·斯特林❷的影

❶ 第一个作品：Maison Delbigot (1968—1973)

❷ 詹姆斯·斯特林 (James Stirling, 1926—1993)，曾在利巴布尔大学学习建筑，该校以巴黎美术学校的教育方针进行教育。到1963年为止与詹姆斯·格文合作设计了哈姆·柯盟公寓 (Ham Common Flats, 伦敦, 1958)，莱斯特大学工学院大楼 (Leicester University Engineering Building, 1963)，之后又设计了剑桥大学历史学院大楼 (1967)等。1971年以后，与麦克·威尔福特 (Michael Wilford) 合作，设计了斯图加特美术馆 (1977)，泰特美术馆扩建工程 (1986)等作品。

formalistes. Et 68 a été important par le fait que ces attitudes-là devenaient quelque chose d'admissible après a l'école. C'est-à-dire, on pouvait proposer quelque chose qui n'était pas dans le cadre, qui était hors du cadre sans être pris pour quelqu'un qui n'est pas sérieux.
Alors, en 68 aussi, je faisais mon premier projet, premier projet destiné à être réalisé : une maison individuelle. Un projet influencé beaucoup par Parent et Virilio d'un côté, mais aussi par James Stirling. Et c'était un projet basé sur trois pentes, comme ça avec des vitrages qui sortaient un peu comme des aquariums. J'ai fait ça avec un de mes amis chez Claude Parent qui devait passer son diplôme en 68.
Alors, en 68, il faut savoir que tout le monde pratiquement a eu son diplôme, sauf moi. C'est pas moi mais mon copain qui présentait son diplôme avec notre projet. Même si vous rendiez rien et si vous expliquiez qu'à cause de la "révolution" vous ne rendiez rien, vous aviez votre diplôme.
Mais là, à partir du moment où je proposais une architecture toute de béton brut qui faisait référence aux études de Parent et Viriliot sur les bunkers, sur la chose militaire, ils m'ont

响。建筑着重表现三个倾斜的面，玻璃面像水族馆那样伸出来。这是和克劳德·巴兰的事务所的一位朋友一起设计的，他在1968年进行了他的毕业设计。

1968年这一年，是校园斗争的一年，谁都能够拿到毕业证书。但是有一个人没有拿到，就是我的这位朋友。虽然他和我一起提交了设计作品，但被退了回来。那个时候是一个非常容易毕业的时代，即使什么都没有提交，只要拿革命做借口就能够得到认可并顺利毕业。我呢，是在几年后提交了别的作品，顺利地拿到了毕业证书。

我们的方案受到巴兰和维利利奥的影响，建筑空间是一个没有垂直墙壁的，完全由倾斜的墙壁组成的空间。当我把要塞和军事设施等作为设计资料进行说明以后，立即就被称为是法西斯主义。在西班牙的弗兰克建有一座专门进行刑讯的监狱，那座监狱为了搅乱人的视觉而没有垂直的墙壁，这被拿出来作为证据。

这个住宅在两三年后得以实现，从建成后的效果来看，完全没有老师们严厉批评的那样用来刑讯的空间形象。相反，主人还非常愉快地住在那里。

那位朋友在两年后再次提交了完全相同的方案，拿到了毕业证书。当年的巴黎美术学院就发生了这么重大的变化。

démontré que ce projet-là était de nature fasciste,et que le fait de mettre les habitants dans des espaces qui étaient basculés, où les murs n'étaient pas verticaux,ça rejoignait les expériences de Franco dans les prisons,qui mettait les prisonniers dans des conditions de torture par des manques de repères spatiaux etc. Cette architecture a été la première construite. Elle a été construite deux ans ou trois ans après. Donc on a pu vérifier que finalement les personnes qui vivent dedans n'ont pas été si torturées que ça. Et je peux même dire qu'elles avaient grand plaisir à vivre dedans,parce que c'est une architecture de plaisir.
Et deux ans après,Roland Baltéra,mon ami,a représenté le même diplôme avec les mêmes documents pour montrer que... et il a eu son diplôme à ce moment-là.
Donc,après j'allais surtout aux Beaux-Arts la nuit,et j'essayais de faire en quelques heures ou dans 2 ou 3 nuits ce que les autres faisaient en trois semaines,parce que j'avais perdu toute foi dans l'Ecole des Beaux-Arts. Qui était en plus dans un grand désordre après Mai 68. Pendant ce temps,j'avais donc construit avec Claude

从克劳德·巴兰那里独立出来

努维尔——那个时候,我只在夜里去美术学院,白天在事务所工作。别的同学花上三个星期做的课题,我只要两三天就能够很好地完成。这样我可以说完全地失去了在美术学院接受教育的欲望。1968年整个体系本身也已经变得支离破碎,因此,我就更感觉不到去大学的意义了。

这期间,我在克劳德·巴兰那里建成了刚才讲过的住宅,同时也作为爱派路那(Epernay)的一个清水混凝土商业中心项目的负责人进行了该工程的工作。

1970年,在事务所工作了5年之后,巴兰对我说:"你已经得到充分的学习了吧。"他告诉我应该自己独立去做了,并把一个小商店的工作给了我,让我从这样规模的东西开始做做看。这个时候,我25岁。因此,我是在学校毕业之前就独立出来创立了事务所,一年之后,我从学校毕业了。

克劳德·巴兰确实给了我许多帮助。他向造型艺术与现代艺术中最重要的展览会——巴黎双年展的总代表乔治·波代尔先生(Georges Boudaille)推荐了我。

Parent ces logements à Neuilly et puis, j'étais devenu chef de projet d'un supermarché fait aussi entièrement en béton brut qui est à Épernay.
Et au bout de 5 ans chez Parent, en 1970, il m'a dit: "Maintenant, tu en sais assez, et il faut que tu t'en ailles. Et je vais te donner des petits projets, par exemple des boutiques aménagées dans des supermarchés, des petites choses que je ne veux pas faire, et tu vas commencer à travailler pour toi."
Donc, j'avais 25 ans, et à 25 ans, j'ai créé mon agence avant d'avoir mon diplôme, puisque j'ai passé mon diplôme un an après. Et là, il est sûr que Claude Parent m'a beaucoup aidé dans ce démarrage, et en particulier en me recommandant à Georges Boudaille, qui était le délégué général de la Biennale de Paris, qui était la principale exposition d'art plastique, d'art contemporain. Et j'ai donc été huit fois l'architecte de cette manifestation, pendant 15 ans, et j'ai fait des aménagements provisoires dans des grands espaces, comme le Parc floral à Paris, le Musée d'Art moderne, Beaubourg après, la Grande Halle de la Vilette.
Et cela me comblait d'aise puisque je retrouvais

❶ 博布鲁（Beaubourg）：1975年举行的第九次巴黎双年展会场，在这块用地上，1977年建起了蓬皮杜文化艺术中心。

在这以后的15年中，我作为双年展的建筑师，有幸进行了巴黎植物园、现代美术馆、博布鲁❶、拉·维莱特的大型艺术展示设施的设计。由于这种关系，我有机会与世界上知名的艺术家相识。又能接触到年轻时被一度关闭的艺术世界，就好像以前的梦想被再次唤醒。

实际上，从开始做建筑的一年后开始，我已经决定再也不画一幅画了。如果要画画的话，不能100%地投入自己的热情和精力，就不可能进入那个世界。那些业余画家或者出于爱好弹小步舞曲和肖邦曲子的人，自己可能会非常欣赏，但其他人绝不会认为好。自己选择了建筑，也就放弃了绘画。

第一个问题就到此为止！（笑）

我认为最好的教育，就是提高自我分析的能力，知道自己现在在做什么

岸田——我想问一下努维尔先生作品的倾向与教育的关系。努维尔先生的作品中有着非常明显的技术形态和工业材料的特征，也有人说具有独特的诗意。刚才您提到从维利利奥先生那里听到各种关于军事设施的

là certaines de mes amours, moi qui voulais devenir peintre, je retrouvais là les contacts avec les grands artistes du monde entier.

Il faut dire qu'un an après être entré en architecture, j'avais décidé de ne plus faire du tout de peinture. Au nom d'une théorie relativement simple peut-être simpliste, c'est que tous ceux qui ont des violons d'Ingres en fait massacrent ce qu'il aiment. Parce que pour faire de la peinture très sérieusement, il faut y mettre tout ce qu'on a, il faut y mettre toute son énergie, tout son savoir, toute sa conviction, c'est une chose qui ne peut pas se faire à moitié. Et quand je vois les peintres du dimanche, ou les gens qui ânonnent leur petit menuet, ou leur petit Chopin sur le piano, à chaque fois, ça écorche les oreilles ou ça écorche l'oeil, ça leur fait peut-être plaisir, mais ça fait pas souvent plaisir aux autres.

Voilà pour la première partie.

SYOGO KISHIDA:

Vos oeuvres sont caractérisées par des formes techniques et des matériaux nouveaux. En même temps on dit qu'elles ont une poésie particulière. Et justement vous avez dit que M. Virilio vous a parlé de beaucoup de choses militaires. Est-ce

谈话，那么，在您所受的教育中，这种影响和现在的作品有很强的关系吗？

努维尔——当然，我想教育的不同会在某种程度上决定发展方向。我认为最好的教育，就是提高自我分析的能力，知道自己现在在做什么，提高自我诊断的能力。

对我来说最难的，就是怎样定位自己的"老师"巴兰和维利利奥。我认为他们是非常优秀的建筑师，但是我明白我自己所要走的路与他们完全不同。有时，他们两人的影响也会出现在我的作品当中。在东京国立剧场的设计竞赛❷中，我做了一个巨大的黑色雕塑造型，一个像巨石一样的作品。许多人都说我受到了他们的影响，我也承认，如果我没有在他们那里学到体量、体块的概念的话，是绝不会做出这样的设计的。

巴兰教给我的，是建筑设计绝不会只有一种设计方法，总会有几种可以选择的方案。在各种各样难以对付的业主面前，首先拿出一个最初的方案，如果他说不行，再拿出第二个、第三个、第四个，这样一次次以不同的形式向同一个业主提出不同的方案，通过研究不同的解决方法，最后必然能够形成一个完成度很高的方案。

❷ 东京国立剧场设计竞赛：1986年日本国建设省举办的第二国立剧场国际设计竞赛，努维尔的方案入围，但没有拿到一等奖。当选的是柳泽孝彦的方案。该建筑已于1996年竣工。

que cette expérience ou alors la formation scolaire vous ont-ils influencé beaucoup?
JEAN NOUVEL:
On est toujours conditionné par son éducation. Je dirais qu'une bonne éducation apprend d'abord à faire son propre diagnostic.
Une chose des plus longues et des plus difficiles a été de me faire une opinion sur d'abord la position de mes "maîtres",de Parent et de Virilio. Et avec toute l'admiration que j'ai pour eux,je savais que ma voie n'était pas la leur. Même s'il m'en reste des choses très fortes qui reviennent à certains moments de façon évidente. Je fais allusion par exemple au projet que j'avais fait pour l'Opéra de Tokyo,un grand monolithe noir. Je crois que je n'aurais jamais pu faire un projet comme ça si je n'avais pas eu cette culture de la masse,de la volumétrie,des choses denses et pleines que j'ai apprise chez eux. Parent m'a aussi appris qu'il n'y a jamais une seule solution en architecture. Je l'ai vu confronté à des clients très très difficiles. Et je l'ai vu proposer un projet,puis un deuxième,puis un troisième,puis un quatrième… Mais ses projets à chaque fois prenaient un autre angle d'attaque et avaient toujours la même qualité.

与此相对，维利利奥教给我的，则是观察力，或者说判断力。在设计时我们要判断的是，当建筑建成的时候，这栋建筑究竟在现实环境中具有什么样的意义，与周围建立了什么样的关系，关于这一点，我受到他的强烈影响。

但是，我在建筑设计中的思考方法与他们有所不同。我最重视"特征"，也就是囊括所有事物的下一维空间，我的设计的基本点就是找出这个特征。因此，在分析阶段，我把特征抽象出来，然后再深入进行方案设计，直到最后我都在处理这种与特征相关的形态问题。我的建筑从形态上来看是非常多样的，其理由也就在这里。我的建筑首先并不是从形式来考虑的，而是在设计的时候，以已有的条件为基础，进行各种各样的分析，导出特征并最后形成一种形式。我最感兴趣的，就是这种战略上的概念。

1970年巴兰患了心脏病，从那以后就很少有接近他的机会了。由于生病的原因，他不再像以前那样对建筑投入大量的精力，结果，我想我从维利利奥那里受到的影响可能更强一些。1970年以后我仍然经常与维利利奥进行交谈，仍然不断地直接受到他本人，或他所写的书的影响。

Quant à Viriliot, il m'a peut-être appris cette notion du diagnostic, cette notion de la distance, de regarder le monde et de chercher les connexions et le sens de tout ce qui se passe autour de la chose construite.

Mais mon approche a été très différente de la leur, puisque moi je suis quelqu'un qui a mis la valeur de spécificité au-dessus de tout. Et qui donc traite le problème formel après avoir intégré le maximum de charges, de données sur le projet. Donc, vous le savez, j'ai fait des projets qui, formellement, sont sur des registres très divers, mais c'est la notion stratégique qui reste la plus importante pour moi.

Ensuite, je dirais que l'expérience avec Parent a été moins prolongée que celle avec Viriliot, parce que Parent a été très malade en 1970 puisqu'il a eu un accident cardiaque, et qu'il a mis moins d'énergie dans son architecture. Par contre, avec Viriliot, j'ai continué à le voir, à parler et à apprendre beaucoup de choses de lui après 70 aussi.

TADAO ANDO:
En ce qui concerne les études d'architecture dans les écoles de notre pays, la théorie d'une part et le projet de l'autre sont clairement divisées. En

建筑师只能自己寻找自己的道路

安藤——在日本，建筑理论与建筑设计是分开的。听到这一段关于克劳德·巴兰与保罗·维利利奥对你的影响的谈话，我想，努维尔先生是理解了建筑理论与设计的关系的。最后，关于建筑理论与设计的关系，以及建筑的学校教育方法，努维尔先生是否有什么能够告诉我们的？

努维尔——向大家提建议是一件非常困难的事情。为什么呢，因为我想，每一个人都必须自己去寻找自己的道路。对于年轻的建筑师来说，最糟糕的情况，就是形成一种文化的定势，或者说一种典型的形式。

对一个建筑单元进行重复设计和生产在历史上曾经非常有意义。在一个世纪以前的欧洲，在建筑领域，非常盛行遵循既定的方法进行重复建设。大家都使用同样的材料，即使在城市规划层面上，大家也都共同遵循一个原则。在当时，已经形成了这样一种整体的建筑体系：就是根据某种类型的规则来建造道路、广场以及建筑。

écoutant votre explication à propos de l'influence que vous avez reçue des M. M. Parent et Virilio, j'ai l'impression de comprendre un peu la relation entre votre théorie et votre création.
Maintenant vous avez ici des étudiants d'architecture, alors donnez-leur quelques conseils à ce propos, s'il vous plaît.
JEAN NOUVEL:
J'ai toujours beaucoup de réticences à donner des conseils. Parce que je crois que chacun doit trouver sa voie, et que je trouve qu'une des pires choses qui se déroule actuellement par rapport aux jeunes architectes, c'est la modélisation culturelle.
Avant, on avait de solides raisons de reproduire des modèles. Il y a encore un siècle, l'architecture était clairement une discipline autonome avec des principes clairement établis. On construisait avec les mêmes matériaux, on était dans des principes urbains qu'on connaissait, on voyait donc très bien comment à partir des typologies classiques on pouvait développer des rues, des places, pour ce qui est de l'Europe. Et au Japon, on voyait tout un art constructif, qui était aussi basé sur le bois, toute une série de choses qui

在日本的木建筑领域，建筑体系与建筑的模数之间也有一个理性的规则，建筑的建造和建成后的生活之间也有一个很强的关系，建筑的世界就是建立在这种秩序的基础上的。

在那个时候，所谓建筑，就是将人工的世界组织到与自然的关系中去，我想当时就是这样一种世界观。但是，从19世纪末到20世纪，这种体系和这种思考方法遭到了全面否定。技术飞速地发展，材料不断地增加，现代主义对既存原理和法则进行了反思，从根本上粉碎了这种观念。城市爆发性地扩张，完全超越了迄今为止的尺度，全世界出现了一种混乱的状态。这已经不是悲叹还是不悲叹的问题了，我们生存下去的条件之一，就是不得不承认这种已经存在的混乱。

因此，与过去的建筑师相比，今天的建筑师必须有意识地运用更多的知识，否则就无法进行建筑师的工作。在19世纪以前，建筑是在学院派的框架之下，只要考虑怎样很好地运用现存的形式，就可以在这种框架下生存下去。

今天，建筑师存在的意义，在于能够对我们所处的这个状态进行深入的分析，提出一个明确的方案。现在，能够对你们讲的，就是要花大量的时间，进行

étaient parfaitement identifiées, parfaitement logiques, parfaitement en liaison avec une vie et avec une production. Et l'architecture avait alors pour claire voie de construire le monde artificiel dans lequel on vivait en relation avec la nature, mais créer ce monde artificiel.

Depuis la fin du 19e siècle et avec l'accélération du 20e siècle, tout cela a explosé. A la fois les techniques, les matériaux ont été multipliés, la modernité a provoqué la remise en cause de tous les principes et de toutes les lois, et le déménagement du territoire, l'explosion urbaine, tout ça a provoqué un immense chaos que tout le monde ne peut que constater. Le problème n'est plus de le déplorer ou de ne pas le déplorer. Ce sont nos conditions de vie aujourd'hui.

Donc je crois que les architectes aujourd'hui sont condamnés à être plus intelligents qu'avant. Puisqu'avant, c'était un jeu académique, il fallait améliorer les modèles existants, quand on faisait de la bonne académie. Maintenant, l'architecte est condamné à l'analyse de chaque situation, et au diagnostic par rapport à chaque situation. Alors, la seule chose que je peux vous dire, c'est prenez les moyens de cette analyse, de cette réflexion, pour pouvoir faire un diagnostic de

广泛深入的思考，这是提出新方案的最重要的一个条件。

仅仅依靠对既存的东西进行复制是无法充分表现出你们的热情的。制造一些仿造的阿尔多·罗西❶、马利奥·博塔❷、安藤忠雄或努维尔等等，没有一点意义。如果从经济或者技术的角度来讲，再生产有一定的意义，但是，作为一个建筑师的作品，尤其是在现代建筑中，对既有作品进行再生产是一件完全没有意义的事情。

我想有一点能够对与文化相关的所有的人来讲：一个文化模式一旦形成，就会形成对此进行简单重复的风潮，因此，我们不能简单地随风而动，每一个人都必须创造出具有正确意义的"真品"。

❶ 阿尔多·罗西（Aldo Rossi，1931—1997）：意大利建筑师。就学于米兰工科大学，从师艾路奈斯·罗杰斯。1966年出版《城市建筑》，1973年与人合著《合理主义建筑》。1969—1971年在米兰工科大学任教授，1972—1974年任瑞士联邦工科大学苏黎世分校（ETHZ）客座教授，1974年又重新回到米兰任教，从1976年开始也在威尼斯建筑大学任教。

❷ 马利奥·博塔（Mario Botta，1943— ）：生于瑞士南部提契诺（Tichno）的曼德利西奥（Mendrisio）。就学于威尼斯建筑大学，1965年进入勒·柯布西埃事务所工作，四年后在路易斯·康手下工作。同年在卢加诺（Lugano）成立了事务所并设计了许多作品。从1983年开始成为瑞士联邦工科大学教授。

年轻的努维尔在工作中

bonne qualité. Et surtout que vos amours ne se traduisent pas par de pâles copies. Parce que combien de faux Aldo Rossi, de faux Mario Botta, de faux Tadao Ando, de faux Jean Nouvel, j'en vois aussi, naissent comme ça en dehors de toute signification. Donc, je crois que ce qu'il y a de pire, c'est cette modélisation culturelle.
On a connu la modélisation technique pour des raisons d'efficacité et de domination de l'économie. Cette modélisation, je dirais, est une honte pour pas mal de politiques je crois, d'hommes politiques; la modélisation culturelle, quand elle correspond à des choses absurdes basées sur ce principe-là, c'est une honte pour les hommes de culture. Cultivez votre authenticité.

ÉTUDIANT:
Dans la première partie de votre discussion, d'après vous, vos parents pensaient que les ingénieurs autant que les enseignants ne soient pas des artistes. Alors pour vous, qu'est-ce que c'est qu'un ingénieurs?

JEAN NOUVEL:
Je fais souvent le parallèle entre l'architecte et le metteur en scène de cinéma. Et je considère que toute production d'un grand bâtiment est

本世纪追求技术表现，
下一个世纪将会隐藏技术

学生——努维尔先生在一开始的谈话中提到，您的父母认为工程师和教师一样，是和艺术没有什么关系的人，我想问一下，现在努维尔先生在自己的实际设计工作中，把工程师这样的人放在什么样的位置上呢？

努维尔——我认为，建筑师的作用，举例来说就很像电影导演，把建筑建起来，就好像制作电影，在制作过程中，各种专业的人参加进来，在各自的位置上发挥自己的作用共同完成一个作品。我是一个建筑师，没有多少兴趣和工程师划分范畴，我认为这没有意义。工程师也好，电影导演也好，都是一起做工作的人。

所谓工程师，也被认为是伟大的建筑师。例如，大家应该知道让·普罗维❶这个人吧，他是工程师还是建筑师呢？还有巴克敏斯特·富勒❷，他又是哪一种人呢？社会上公认的职业名称与这个人做的事情连在一起，我想没有多少意义。在制造电影或建筑的时候所必需的，是一个导演。建筑和电影一样，是以工作组

❶ 让·普罗维（Jean Prouve，1901—1984）：接受了工程师的教育之后，于1917—1920年在艾米路·罗波特（Emile Robert）手下学习铸铁的制造技术，1918年制作出第一个作品，1923年成立制作工房，最初的工作是制作一些铸铁格子等东西，之后开始制作金属构筑物，并成为装配化施工方法的先驱者，1939年之后建造了医院、学校、工厂、住宅等建筑。第二次世界大战之后，他把大量精力投入到住宅的装配化生产中。普罗维一贯将结构构件与围合构件相分离，在"罗兰·卡罗斯航空俱乐部"的设计中首次将幕墙的设计方法运用到实际工程中。

❷ 巴克敏斯特·富勒（Buckminster Fuller，1895—1983）：虽然就学于哈佛大学，但对学院派教育没有兴趣。第一次世界大战时参加海军，第二次应用工学和作战，这影响到他后来的设计思想。他的结构技术研究的成果主要表现在用四面体和八面体构成的"吉奥德西克·穹顶"（Geodesic Domes），这栋建筑曾作为"联合坦克修理工厂"、"蒙特利尔万博美国馆"等。1949—1975年在南伊利诺州工科大学任教授。著作有《月亮的九把锁》(1938年)、《探索思考的几何学》(与苹果白合著，纽约，1975）等。

un peu l'équivalent d'une grande production cinématographique. Ce que je trouve formidable au cinéma, c'est que dans une division du travail bien construit, chacun a trouvé sa noblesse et revendique la précision de son travail.
Les termes architecte, ingénieur pour moi ont peu d'importance. Un ingénieur peut être un réalisateur de cinéma au sens où je l'entends, il peut être à l'origine d'un projet.
On a vu de grands ingénieurs qui étaient de grands architectes. On a vu des architectes être aussi des ingénieurs, et on ne sait plus vraiment ce qu'ils sont. Est-ce que Prouvé était ingénieur ou architecte? Est-ce que Buckmister Fuller était ingénieur ou architecte?
La seule que je sais, c'est qu'à chaque fois qu'on fait un film, qu'on fait un bâtiment, il faut qu'il y ait un réalisateur, et il faut que la répartition des rôles soit assumée en toute clarté et en toute légitimité et en tout honneur que chacun revendique. En France, on vit quelquefois des jalousies comme ça où les ingénieurs essaient de parfaire ce que veut l'architecte. Des bagarres entre les ingénieurs et les architectes, ce sont des choses sans intérêt. Cela vient souvent du fait que ce sont des mariages forcés: pour des raisons

的形式共同作业的，如果每个人都能认真地完成自己的任务就会产生好的作品，就是这么单纯的事情。有时，工程师与建筑师之间也会存在嫉妒的事情，工程师不会这些、建筑师不会那些等等，这些互相中伤的议论都是完全没有意义的事情。

当然，在实际建造的过程中，建筑师和工程师的合作是必不可少的。我的作品也是这样，如果离开了他们的合作是不可能建成的。我自己会选择工程师，相反工程师也会来选择我，不管怎样，我的理念是：我只选择那些可以愉快地合作的工程师。例如，对我来说如果是一个重要的工程，而且自己能够选择工程师的话，我会选择与英国的托尼·费茨巴德利克和保罗·纽塔❸一起工作。和他们合作是最有趣的，他们可以说是我最好的伙伴。

❸ 托尼·费茨巴德利克（Tony Fitzpatrick）和保罗·纽塔（Paul Nutall）：1964年在英国成立的结构设计事务所奥布·阿拉部的成员。和努维尔一起进行了"卡梯尔基金会"的设计工作。

我想，20世纪的重点，还是应该在技术方面。事实上工程师们将技术进行展现，并从中获得了成就。而现在，这种状况完全改变了，现在这个时代的感觉，不是将技术展现出来，而是让人不知道建筑是怎样建成的，这样的技术才会给人更新的感觉，恐怕无人知晓从本世纪末到下一个世纪，究竟会形成什么样的一个东西。但如果建筑形成的过程让人无法从视觉上认识出来，这样的建筑师或者工程师肯定可以被称为是

économiques,il y a deux personnes qui arrivent ensemble,il y a des querelles de pouvoir.
Moi je sais que beaucoup de mes projets,je ne peux les faire qu'avec les ingénieurs que je respecte,que j'aime et que je choisis. Et qui me choisissent. Mais quand un projet est très important et que j'ai la liberté souvent,je travaille avec Tony Fitzpatrick et Paul Nutall,qui sont, je dirais,mes compères favoris dans le domaine de la collaboration ingénieurarchitecte.
Il y a autre chose qui a beaucoup changé,c'est qu'on peut dire que dans la poétique de ce siècle et dans son mythe,il y a eu la technique. Et à partir de là,les ingénieurs revendiquent une sorte de mise en scène,d'expression de la technique. Mais actuellement,tout se renverse,c'est-à-dire que la chose la plus extraordinaire,c'est celle qui atteint un certain résultat sans qu'on voit comment on atteint ce résultat. Le grand ingénieur de cette fin de siècle ou du siècle prochain,ou les grands architectes,sont ceux qui arriveront à faire oublier la technique. Pour cela il faut être virtuose.

ÉTUDIANT:
Est-ce que vous avez déjà collaboré avec d'autres architectes?

最前沿的。虽然说只有天才才能做出来。

即使在合作的情况下，建筑师也和电影导演一样，最好是一个人做

学生——至今为止，您有没有和其他的建筑师一起合作进行工作？

努维尔——就像我刚才讲的那样，我相信每一个项目中只能有一个导演，不可能有两个或者三个导演。不过，如果把工作明确地分配好的话，我想合作的形式是经常会有的。

迄今为止，我也有联名进行一个项目的经历，但不作为导演的情况一次也没有。共同设计的时候虽然会有好几个人，但那并不是像作家那样对作品进行分工，他们只是经营方面的合作者，或者在技术方面对我进行辅助，有过这样的一些共同设计。

在设计"阿拉伯世界研究所"的时候，我和一些年轻的建筑师们共同使用过同一个场所。那个时候因为穷，大家共同租了一个大的场所，大家在同一个地方做着各自不同的设计。你们知道，"阿拉伯世界研究所"是一个指名竞赛，被指名的是我个人，但我邀请

JEAN NOUVEL:
Je viens de dire que je crois que dans un projet il faut toujours qu'il y ait un réalisateur. Je ne crois pas aux projets avec deux réalisateurs,ou trois réalisateurs. Par contre,je crois à des formes de collaboration,dans une claire division du travail.
Alors,il m'est arrivé de co-signer des projets,il ne m'est jamais arrivé de ne pas tenir le rôle de réalisateur. Et dans l'essentiel des cas,ces projets co-signés correspondent à des partenaires qui avaient comme mission par exemple la structuration ou la gestion de mon agence,ou qui avaient je dirais,dans notre société,des missions complémentaires aux miennes. Ou commercial,ou technique ou économique.
Alors il y a d'autres cas de figure. Il y avait un cas comme l'Institut du Monde Arabe,où des jeunes architectes dans un même local avaient décidé de partager une ressource économique. Donc,quand j'ai été invité sur le concours de l'Institut du Monde Arabe,j'ai proposé à mes associés du moment et à Architecture Studio de me rejoindre pour partager un peu quelque chose qui arrivait. Mais j'ai toujours gardé le contrôle au plan justement de cette réalisation.

了在同一个场所进行工作的伙伴,一个名叫"建筑工作室"❶的小组,我们一起参加了这个竞赛。在这种情况下,我仍然坚持一定要作为项目负责人。

安藤——今天您给我们讲了这么长时间,真是非常感谢。我第一次见到努维尔先生可能是在1981年,我们的交往是从在巴黎的工作室见面以后开始的。从那以后您一直都非常活跃。

前一段时间我看了您设计的柏林百货公司❷。柏林规划这样一个项目,很多人都参加了工作,但从整体上来看并不是很成功,或许努维尔先生也很清楚。在这幢建筑里面,您越过柏林沉重的历史,使用了玻璃这样非常轻的材料,向我们传达了强烈的信息,这一点让人很有感触。

巴黎的卡梯尔基金会画廊建筑❸也是这样。前几天我从三宅一生先生那里听说,今秋的展览会就在那里开,他也从建筑中感受到很大的震动。从这些方面来讲,我相信努维尔是现代建筑师中对我们触动最大的建筑师。

❶ 建筑工作室:1973年以马丁·罗班(Martin Robain)为首成立的共有六人的设计事务所。主要作品有"未来高中"(1987)、"欧洲议会大厦"(1998)等。

❷ 柏林百货公司:在1991年的指名设计竞赛中,努维尔的方案当选。建筑用地设在原东柏林的历史街区中,努维尔的方案是在砖砌的厚重风格的街区中插入一个玻璃幕墙的透明盒子,使内部的活力渗透到建筑外部。另外,在形体的上下左右设计了大小不一的圆锥形或圆柱形空间,引入自然光线,并形成视线的交错和影响的互映。尤其是在中心处设计了一个半径35米的倒圆锥的玻璃体,目的是体现出以往百货店里常见的向心空间。

❸ 卡梯尔基金会画廊(1994):卡梯尔基金会的办公室和展示空间。设计有诸多要求,例如保存基地树木。在原有的建筑遗址上进行建造,新建筑的高度也受到限制,不得高出街道两侧的建筑等。在这种情况下,努维尔在临街的一侧设计了一片高18米的玻璃幕墙,在用地的内部又设计了一个由两片玻璃幕墙相夹的高31米的空间。三片玻璃幕墙映射出人的活动和周围的树木,并将内部空间变得宽敞,该设计有意模糊了内外的界限。

TADAO ANDO:
Merci beaucoup.
Je vous ai rencontré en 1981 à votre agence de Paris. Depuis lors, je pense que vos oeuvres se sont bien développées; par exemple, récemment j'ai vu votre grand magasin à Berlin. Dans le contexte historique très "lourd" de Berlin, vous avez mis des matériaux légers, comme le verre et j'ai été très impressionné par son message fort. En outre, le bâtiment de Cartier à Paris est tellement présent que M. Issey Miyake, en organisant son exposition pour l'automne, a été très inspiré. En somme, je crois que parmi les autres architectes contemporains, vous donnez un des plus forts messages dans notre domaine. Merci encore pour votre conférence.

1998年6月29日 （星期一） 10:30am——12:00am 主持：铃木博之

理卡多·雷可瑞塔
RICARDO LEGORRETA

理卡多·雷可瑞塔于1937年生于墨西哥城的一个银行家家庭，他在墨西哥大学建筑学专业学习的时候，就在杰赛·比拉格兰的手下积累着建筑设计的实际经验。大学毕业两年之后，他就成为比拉格兰的合伙人，并于1963年独立开业。

雷可瑞塔追求对墨西哥传统、历史、风土、民主性等本土形象的强烈表达，同时也采用抽象的现代建筑手法。在他的建筑中，有着独具特色的光和色的处理，也有着用水来展示的空间，还有粗糙质感的大面积的墙体。在这些语汇组成的连续的空间中，我们真切地读出了延续了几千年的墨西哥文化。值得注意的是，他在设计中避免了对传统和乡土的东西的直接模仿，他的空间甚至可以说是完全通过抽象来支配的。

同时，从现代建筑的观点来看，雷可瑞塔的设计思想也是自由的。他虽然被现代建筑中的抽象美学概念深深吸引，但他所追求的理想却与现代建筑完全不同。他热爱本土文化，并努力使之变得完美和理想化。

20世纪60年代，一场反思现代建筑形式的思潮席卷世界，使得各国都开始重新考虑地域传统，并产生了"地域主义"。雷可瑞塔也常常被纳入"地域主义"这个流派之中。但是，我们不应该忘记，他不仅在墨西哥国内创作了"卡米诺·雷阿宾馆"（墨西哥城/1968，伊克斯塔巴/1981）、雷诺工厂（戈麦·帕拉西亚/1985）等一系列作品，同时也将活动范围扩展到了美国以及世界各地。

即使在"地域主义"思想的创造力开始减弱的今天，他设计的建筑仍然没有失去其生命力。

（驹田刚司）

墨西哥城卡米诺·雷阿宾馆 (CAMINO REAL HOTEL MEXICO Mexico City, Mexico) /1968
Photographer Lourdes Legorreta

美国洛杉矶Pershing 广场(PERSHING SQUARE Los Angeles, CA, USA.) /1993
Photographer Lourdes Legorreta

热爱墨西哥的父亲带我
从小就在国内四处旅游

铃木博之——今天我们希望能够了解墨西哥建筑师雷可瑞塔成长的背景和所接受的教育。

您是在墨西哥大学接受的建筑教育,那么大学以前,您是在什么地方上的中学,中学教育情况又是怎样的呢?开始学习建筑之前的环境对您的专业工作具有什么影响和意义呢?我们想听一听这些事情。

理卡多·雷可瑞塔——我曾经接受过两类教育:一是通过一种非常传统的方式,也就是先上私立学校(小学),接着再上中学。这是一个非常简单的过程,这中间不存在任何复杂的地方。另一种与之平行的教育方式则是非正式的。它包括了……我父亲非常热爱墨西哥,因此在我很小很小的时候,他就带我遍游了许多的地方。我们游历了墨西哥的许多小村落、城市和农场,通过这个过程,我开始熟悉建筑。我也很习惯于在那些空间中生活,习惯于以一种非正式的方式对它们进行观察和学习。

有意思的是,我的家族中并不存在任何建筑和艺术的背景❶。我的家庭从事的是金融方面的工作。我已

雷可瑞塔幼年时

❶ 雷可瑞塔1931年出生于一个西班牙后裔的望族家庭,他的父亲对其灌输过强烈的社会责任和纪律教育,其家庭日常生活与宗教的密切相关性,培养了他对建筑精神层面的热爱。——译者注

HIROYUKI SUZUKI:
What kind of education did you have before you start learning architecture?
RICARDO LEGORRETA:
I think that I had two kinds of education: one was from the very traditional way, which was to go to a school-private school(primary) and then what we call "secondary". So that was very simple, without any complications. The other one that I could say went parallel was the education that was nothing official. This consist... my father loved Mexico very much, so he took me around the country when I was very, very little. So I got used to visiting all the places. By visiting small villages, cities, and ranches in Mexico, I got used to architecture. I got used to live in those spaces, to observe, and without doing it officially, to study them. Something curious was that in my family, there was no background of architecture or art. My family was on the financing world. So I don't remember when I decided to be an architect. There was no special decision in my life about being an architect. It just came very natural. May be it was the

经不记得自己是在什么时候决定去作一名建筑师的。这对我来说,并不是一个非常特殊的决定,一切都发生得非常自然。也许这只是我对自己根本不喜欢的那个金融行业,所做出的一种自然反应。

在大学里的学习,
有幸遇到醉心于建筑的建筑师们

第二个阶段就是我准备去上大学的那段日子,我希望在接受大学教育的同时,也能直接接触一些实际的建筑工程。所以当有一天,我看到一幢建筑事务所的大楼,我很喜欢这个建筑,它显得非常完美而且结构清晰,于是,我便决定要去设计这幢建筑的建筑师事务所去找工作。

将学校的教育和建筑师的言传身教相结合的学习方式使我获益匪浅。我在墨西哥学习的五年,实际上包含了三个不同方面的内容,它们分别来自学校、建筑师以及我在全国各地的旅行考察。这些帮助我认识到,如果要研究你所在的国家……你就要接受各方面的影响,但是不要去模仿或者设计一些糟糕的,或者和过去的建筑相似的东西。你要以自身的能力去立足于你所处的时代。

logical reaction to the financing world which I don't like at all.
The second phase was when I was ready to go to the university, and I decided that I wanted to have at the same time the education of the university and the education of a direct communication with the architect working. So one day, I saw a building which I like and was very finished and constructive, so I decided to go to the office of that architect and ask for work.
So, the combination of the school and being close to the architect was very good for me. So the five years of university-in Mexico there are five years-went with the three aspects of education; the school, the architect, and the visits and travelling around my country. That helped me very much to understand that to study your country is… you receive the influence without going into copying or making bad or things that look old. You want to be on your time but with your 'roots'.
Another special thing of this period was that the architect I worked with was typical architect of the modern movement and I didn't

在此期间还有一点特别的是，我所为之工作的那位建筑师是一位现代建筑运动中的典型代表❶。虽然我并不太喜欢这样的建筑，但是我从他身上学到了建筑设计的基本原则和规范，以及对建筑的热爱。我为他工作了12年，在我离开他的事务所的时候，我已经决定将自己余生所有的力量，投入到自己的建筑事业中。

格罗皮乌斯的忠告："尽量多去旅游"

就在我完成学业的时候，一件偶然的事件对我产生了重大的影响。你们可能会不记得他的名字了，但这是一位非常著名的德国教师———瓦尔特·格罗皮乌斯❷，著名的包豪斯❸学校的校长。他来墨西哥访问，并在旅馆中举行了一个聚会。我去了那里，并请厨房里的人帮助，使我穿过厨房到达举行聚会的地方。我请求他给我一些建议，告诉我应该怎样去学习建筑，该做些什么，怎么做，从而进一步学习建筑。他问我是否准备留在墨西哥生活和工作，我说："是的。"因此他对我说："那就住在墨西哥，然后尽可能多地去旅行。"所以从那时候起，我就开始攒钱，然后到处旅行，

❶ 雷可瑞塔大学毕业后进入卡西尔（J. Villgram Carcia）设计事务所工作，卡西尔是当时致力于发展墨西哥现代建筑传统的前辈之一。
——译者注

❷ 瓦尔特·格罗皮乌斯（1883—1969）：现代建筑的传道师，在理论和实践中探索在现代工业技术前提下的建筑设计。当时雷可瑞塔二十二三岁，格罗皮乌斯七十多岁。

❸ 包豪斯：1919年设立于德国魏玛市（Weimar，后迁移至德绍（Dessau））的造型艺术学校。目标是将包括建筑在内的各种造型艺术与现代工学相结合。该校的第一任校长是格罗皮乌斯，1933年被纳粹关闭。

like the architecture of that movement too much. But I learned the discipline, the ethics, and the love for architecture from this architect. I worked him for 12 years, so when I left the office I was ready to apply all the influences of the rest of my life, and started to do my own architecture.
At that time, when I finished school, I got a very strong influence by just a circumstance. You may not remember the name, but there was a very famous German teacher, Walter Gropius which was the head of that famous school of the Bauhaus. He went in a visit to Mexico and he had a party in a hotel. So I went and I asked for a favor to the people in the kitchen to let me go through the kitchen to where the party was. And I asked him for his advice of what to do and how to keep learning architecture. He asked me if I was planning to live and work in Mexico, and I said "Yes". So he said to me, "Stay in Mexico and travel as much as you can ." So since then, I save money to travel as much as I can, and I keep learning and learning. So trying to resume me, my life as a student was quite

并且不断地学习。我的一生就像是个学生，不断地更新自己，这对我来说是很正常、很自然的事情。我很幸运，总能受到那些非常热爱这个专业的人的影响。

铃木博之——我们已经明白了雷可瑞塔先生的教育背景，以及怎样开始学习建筑的情况。我觉得您学习建筑的方法是非常好的。这种方法在墨西哥的学生当中是一种典型的方法呢，还是一个特殊的例子呢？另外，我们现在正在大学里教书，我们想知道，对雷可瑞塔先生来讲，大学对您具有什么意义呢？

理卡多·雷可瑞塔——从某种程度上来说，和教授保持接触是一种非常典型的学习方式。所不同的是各个学生在其中付出的努力，我就是通过和老师保持亲密联系这样的方法来学习的。

从墨西哥的街道和乡村中学习

铃木博之——从雷可瑞塔先生的作品中我们可以看到非常大胆的色彩和形式处理，以及独特的诗情画意。雷可瑞塔先生受到现代建筑大师格罗皮乌斯的引导，而且从师十二年的建筑师也是一位现代主义建筑师。您接触的这些建筑师都是崇尚现代风格的建筑师，而

normal. But I was very lucky to be influenced by people who loved the profession.
HIROYUKI SUZUKI:
Is that a typical way of learning architecture in Mexico?
RICARDO LEGORRETA:
In one way, it's a typical way in the sense of having the contact with the professor. What is not common is the effort of the student, and that was my case-of getting very close to a teacher.
HIROYUKI SUZUKI:
How did modern and Mexican tradition come together in your architecture?
RICARDO LEGORRETA:
It came together, really in a very special way, because I receive from these architects, the advice of Mr. Gropius, the Professors of the university, I received the education of knowing what was architecture, but they didn't really interest me regarding the concept of design. The concept of the colors, proportions, using the walls, all that I learned from the towns and the villages of Mexico. So they came together in a very unusual way, because on one side the

the picture in the middle
Photographer Lourdes Legorreta

LEGORRETA + LEGORRETA 办公室
(LEGORRETA + LEGORRETA OFFICE Mexico City, Mexico) /1966

墨西哥坎昆岛的卡米诺·雷阿宾
(CAMINO REAL HOTEL CANCU
Cancun, Mexico) /1975
Archive Legorreta + Legorreta

美国得克萨斯西湖公园主题规:
(合作者: Mitchell, Giurgola Bart
Myers and Peter Walker , Martl
Schwartz) (WESTLAKE PAF
MASTER PLAN Westlake
Southlake, Dallas, TX, USA)
1985
Photographer Lourdes Legorreta

尼加拉瓜都会大教堂
(METROPOLITAN CATHEDRAL C
MANAGUA Managua, Nicaragua)
1993
Photographer Lourdes Legorreta

在您的作品中我们却看到墨西哥式的传统建筑表现，您是怎样把两者统一起来的呢？您能否谈一下这方面的事情。

理卡多·雷可瑞塔——它们两者确实是以一种非常特殊的方式结合起来的，因为我受到这样一些建筑师的影响，例如格罗皮乌斯先生的建议，大学里的教授，学校教育也告诉过我建筑是什么，但是这些并没有激发起我自身对于设计概念的认识。我的那些关于色彩、比例的概念，那些关于墙体运用的知识，都来自于墨西哥的城市和村庄。它们之间的结合是以一种很不寻常的方式完成的。因为一方面，那些来自于墨西哥城市、村庄的影响是非常"感性"的；而另一方面，我也具有关于规范和专业概念方面的知识。因此我所设计的建筑应该既是属于我们这个时代的，同时也是坚定地植根于我所生长的国家的，我试着用这样的方式来实现两者的结合。

举两个例子来说明，例如对色彩❶的运用，正如我所说，是完全感性的。我所看到的那些村民运用色彩的方式，也就是我的色彩运用方式。我所能够告诉你们的就是，这里面没有涉及到任何理性方法。可以说，我对于色彩运用几乎不受自己的控制。

另一个例子是和光有关的。我向那些根据房间内

❶ 色彩：墨西哥建筑的特征之一。这个讲座之后，他把自己做的剪纸放到我们面前，这是一个用橘黄色和紫色的纸非常精巧地刻出来的作品，这种颜色的使用方法让人想起他的建筑，也让人感受到他细腻的情怀，这是一个非常可爱的礼物。

influence from the towns and the villages of Mexico is purely 'emotional'. On the other hand, I had the discipline and the concept of the profession. So, I was able to combine them and in that way, tried to do architecture that belongs to our time but has that very strong root of my country.

So just to mention two examples, the use of color, as I said, is purely emotional. And the way I see people in the villages——they use color——is the way I use it, and the only thing I can tell you is that there is no intellectual approach. I could say that we're almost 'irresponsible' in the use of color. The other example is relating to light. I learned from these people that they open their window considering the life inside of the building, not from the outside, or the elevation. They open their window to receive the sun at the certain hour of the day, or to receive the light in a completely different way. So, I design much more from the interior than from the exterior.

HIROYUKI SUZUKI:

Did you feel like you were going against the current of the modern movement?

生活的需要，而不是根据房间外，或是建筑立面需要来设置窗户的人学习。这些人打开窗户是为了在每天的特定时间迎接阳光，或者是以一种完全不同的方式引入光线。所以我的设计更多是从内部出发，而不是外部。

我内心中的浪漫主义思想使我无法接受强硬的形式

铃木博之——这种从墨西哥的风土和传统中学习建筑的思考方法意味深远。其实在雷可瑞塔先生学习结束，考虑在建筑中更加积极地反映墨西哥风土人情的时候，正好是世界范围内现代主义运动推广发展的时期。在这种战后现代主义建筑的发展高潮中，您自己是否意识到您选择了一条独特的道路？

理卡多·雷可瑞塔——根据我对建筑设计的理解，这算不上是一个明确的决定。我认为最重要的是要非常诚实地对待你所面临的问题和你自己，然后尽可能地去做好。也就是说我从来没想过要发展出一种特定的建筑风格。我不太喜欢那种国际式建筑冰冷、机械的风格，但也仅限于此。我反对那种僵化的建筑设计方法，而倾向于使用一种更富于感情的方式去表达。我必须说我是一个特别浪漫的人，而浪漫主义者和教条

RICARDO LEGORRETA:
It was not a very specific decision about doing architecture the way I think. I believe that the most important thing is to be very honest with the problem, with yourself, and then do the best you can. So, I never had the idea of developing a 'style' of architecture. I had a little bit of the reaction to the cold and industrial aspect of the international style, but that was all. It was more of a reaction against a very rigid way of doing architecture and the love for the emotion of it. I must say I am an extremely romantic person. The romanticist with the rigidity doesn't go together.
HIROYUKI SUZUKI:
What do you think of the influence that journalism give to the students?
RICARDO LEGORRETA:
I think that's a very, very important question and a very important subject for the students. On one side, we all know travelling is getting more expensive and more complicated every day. On the other hand, the real way to judge architecture is to be 'inside' of the architecture, to 'visit' the

主义者是不可能走到一起的。

铃木博之——明白了。

用自己的眼来看，用自己的身体来感受

铃木博之——我觉得，对学生来讲，与旅行相比，现在的各种报道、文章具有更大的影响力。您从学生时代开始到现在，和报道、文章之间处于一种什么样的关系？或者说，对现在的学生来讲，应该怎样看待这些从报道和文章中得来的信息呢？

理卡多·雷可瑞塔——我认为对于一个学生来说，这不仅是一个非常、非常重要的问题，同时也是一个相当重要的课题。一方面，我们都知道现在旅行变得越来越昂贵和复杂了。但是另一方面，要评判一幢建筑的真正方法是进入建筑的内部，去亲自走访这幢建筑。杂志实际上只是一种将建筑信息传递给学生的工具。我不会像阿尔多·罗西❶那样极端，他曾告诉学生："请不要看那些杂志。"但是我会建议，一定要非常谨慎地对待那些来自于杂志、录像、网络上的知识，因为这些很可能会使你仅仅从外观或表面去评价一幢建筑。

我想我可以用一个你们都知道的事例来回答这个

❶ 阿尔多·罗西 (1931—1997)：对欧洲传统城市与建筑中潜在的空间原型进行研究，探索其在现代城市中的应用。其个人经历请参照69页注1。

architecture. So, journalism is in fact a tool of information for students on architecture. I will not go to the extreme of Aldo Rossi when he said to his students, "Please don't look at magazines ." But, I will advise you to study very carefully what you see on the magazines and videos and internet, because the risk is that you can end up judging architecture by the appearance or the skin only.

I think I can answer or complete my answer by an example that you all know. Certain—I don't say you all, but certain students in Mexico think that the essence of Ando's architecture is concrete. And it's a big mistake to say that. The essence of Ando's architecture has nothing to do with concrete. The concrete is an element of expression but it has nothing to do with it. He says his essence is spirituality, is proportion, is light, and of course is a great talent.

So you have to be very careful and spend enough time studying the magazines, or films, or videos in order to really understand and read what the architect has to say, and

问题，或者说使我的回答更完整。当然我这里并不是指你们中的每个人，但是确实有一些墨西哥学生认为安藤建筑的精髓在于混凝土。这里面存在一个很大的误解。安藤建筑的精髓与混凝土并没有关系，混凝土其实只是一种表达元素，而与安藤建筑的实质并没有关系。安藤曾说过，他的建筑实质是一种精神，是比例，是光线，当然也是一种伟大的天才。

所以你们必须非常谨慎，要花足够的时间去研究那些杂志、电影、录像，这样才能够真正地理解和明白这些建筑师所要表达的东西，他们工作的地点，设计的文化背景、体系和他们所思考的内容，这样才能够理解它。

你们应该将这些杂志视作一种工具，而不是生活的目标。再回到刚才的例子，安藤的作品之所以会被出版是因为他的作品是非常优秀的，但是反过来也可能是这样的，我们可能也会认为，正因为他的作品被出版了，所以才是优秀的。

所以不要因为旅行是困难而昂贵的，就忘记了旅行所带来的好处。今天上午我们还在讨论，我希望能够提供帮助，通过与你们大学合作，鼓励日本和墨西哥之间的学生互访。当安藤先生去墨西哥的时候，我和他将就某些问题确认一下，我希望你们也能去那里。

铃木博之——从刚才起您就一直在强调旅行的重要性，

where he works, in what culture, what system and what is his thinking. Then, you can understand it.
So, look at journalism as a tool, not as the objective of your life. Again, going to the example, Ando is published because he's very good and the opposite could be that we could think that he's good because he's published.
So don't forget, even if it is difficult and expensive, don't forget the benefit of travelling. And we were talking this morning and I hope and I offer my help, I hope we can do that with your university, which is an interchange of students in some way between Japan and Mexico. I have had problems trying to convince Mr. Ando in going to Mexico, where I hope you'll go.
HIROYUKI SUZUKI:
Can you tell us what kind of place we should visit?
RICARDO LEGORRETA:
I think each one of us, we know what we like. You can find yourself what is your tendency, where you're happy, where you feel very well. When you apply this to

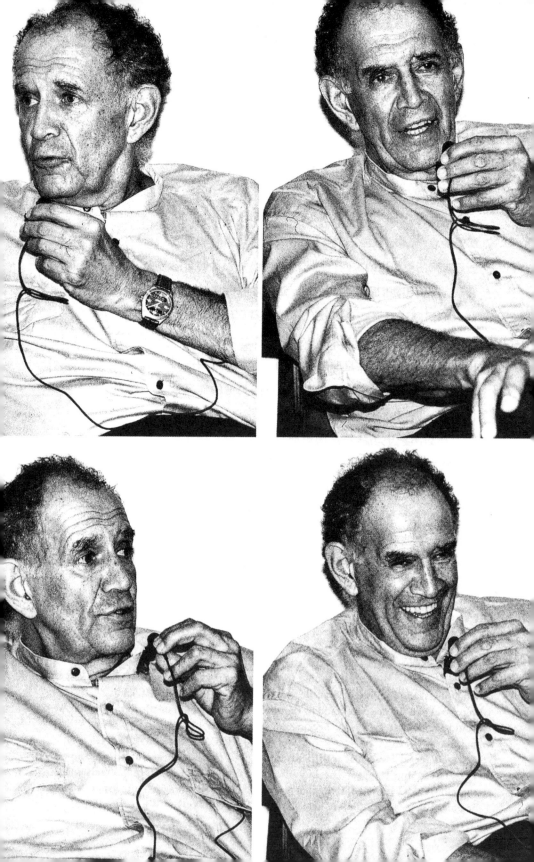

您能否指点一下我们在座的学生,应该进行什么样的旅行、应该怎样进行旅行?

理卡多·雷可瑞塔——我想我们每个人都清楚自己所喜欢的东西。你可以发现你自己的兴趣所在,那些使你愉快的地方,那些让你感觉很棒的地方。如果将这些感觉运用到建筑上,我们就会知道自己喜欢怎样一种空间。从这个角度来说,这些地方对你来说就是很重要、需要去探访的地方。而另一方面,有些建筑和空间会让我们产生惊奇的感觉,这些也是去访问的好地方。因为我们对它们不了解,而这样的探访将帮助我们去理解这些我们并不了解的东西。

我们之所以这样做是因为,世界正在以这样的方式敞开它的大门,而我们可以走出自己的国家去工作。所以我们必须接受挑战,去理解其他文化,同时又要立足自己的国家,强调自己的个性。出于这样的原因,我要提醒你,我非常急切地想知道安藤将如何完成在得克萨斯的沃斯堡美术馆❶。在这个项目中,我倾注了我的所有热情,而我要不无遗憾地说,输给安藤是一件令人骄傲的事。同时我也非常有兴趣知道,你们怎么看待美国南部得克萨斯的空间,那里是如此的空旷寂寥,无边无际。

我对封闭空间的反映,与安藤对开敞空间的反映

❶ 得克萨斯州沃斯堡美术馆:与路易斯·康设计的金贝尔美术馆相邻的一座现代美术馆。1997年举行了该馆的指名设计竞赛,安藤忠雄的方案当选。

architecture, we know what are the kind of the spaces we like. And those are the ones I think is important to go and visit, in one side. On the other hand, there are spaces or architecture that 'surprise' us. That also is good to visit, because we don't understand them and is very good go and see something which you do not understand.

The reason for this is that the world is opening doors in such a way that we can work outside of our countries. So, we have to accept the challenge of understanding other cultures and at the same time maintain our own roots and our own personality. In this case, I want to mention to you that I'm anxious to see what Ando is going to do in Texas in the Museum of Fort Worth, in which I put all my enthusiasm and I'm not happy to say I'm honored to have lost to Ando. And it was very interesting to see how you see the spaces of the South of the United States of Texas, which are empty, almost endless.

How I reacted to enclose the space and how Ando reacted to open the space to the outside was completely the opposite. This is

87

是完全相反的两个极端。当我们面对另一种文化的时候，这种吸引人的地方就体现了建筑的价值观。

以旅行者的眼光来看日本

铃木博之——雷可瑞塔先生已经先后五次来到日本，刚才也对安藤先生的建筑进行了深入的分析。日本的传统建筑和现代城市都拥有多样的层面，那么您是怎么看待日本传统建筑与现代城市的关系呢？

理卡多·雷可瑞塔——我个人对于日本了解不够，很难清晰回答你的问题，但是我可以感觉到，在传统的日本城市建筑和当代的日本建筑之间存在着强烈的联系。我总是感觉到，你们对于两者都是认同的。例如，也许你未曾感觉到，但是作为一个旅行者，我在正在建设的当代日本建筑中，发现了一种典型的日本文化特征。当然我们不能将两者作一种直接的对比。它们的需要、环境和生活方式是完全不同的。

我所能告诉你的是，我清楚地感觉到你们正在解决——我所说的是那些优秀的日本建筑师——像日本这样的国家的问题，但是你们并不打算将这种解决方案强加给社会。

the value of architecture what is fascinating about going to another cultures.
HIROYUKI SUZUKI:
What do you think of Japanese traditional architecture and modern cities?
RICARDO LEGORRETA:
Well, I don't know enough about Japan to give a very clear answer, but I can tell you that I feel very strong relations between the traditional Japanese cities and architecture, and what architecture is doing in Japan now. There's always a sense that you identify in both. For example, maybe you don't feel it, but as a visitor I see a tremendous character of Japanese culture in the contemporary architecture you're doing. And of course, we cannot compare directly one with the other. The needs, the circumstances, the way of living, is completely different.
I can tell you that there is that very clear feeling that you're solving—I'm talking about you, the good architects from Japan—you're solving the problems of a country like Japan, but you're not trying to impose solutions to the society.

认识墨西哥本土文化与西班牙文化的共性,享受两者之间的矛盾

铃木博之——在您的建筑中我们看到非常强烈的色彩和坚固的形式,但是我们也感觉到一种从现代主义的单一性和纯粹性中解脱出来的自由。

今天您给我们谈了旅行的话题,也谈了安藤先生的建筑和您的建筑之间不同的思考方法。其实,旅行或者体验各种事物的时候,就是我们遭遇相反的或者是相矛盾的东西的时候。雷可瑞塔先生的建筑并不是在整理矛盾,而是用丰富的形体去容纳那些对立的事物,使之共存。现在能否请您谈一下有关这种自由的价值观的秘密。

理卡多·雷可瑞塔——从这方面来说我想我们墨西哥文化是很幸运的,因为我们结合了印第安和西班牙这两种文化❶,而西班牙文化同时又受到了伊斯兰文化的影响。所以这两种文化从某些方面来说是有着很大矛盾的,但同时你也能够发现,两者之间在某种程度上还有一些巧合的相似。例如,偏爱非常强烈的形体和很宽敞的空间。这两种文化都重视对建筑墙体和色彩的运用。我发现能够对我的设计有所帮助的方法,

❶ 印第安文化和西班牙文化:印第安有着悠久的传统文化,其历史可以追溯至公元前2世纪,从16世纪西班牙征服了墨西哥以后,大量的外来文化和土著文化共存,也许正像雷可瑞塔讲的那样,能够接受矛盾的这种文化的土壤,造就了巴拉甘和雷可瑞塔等人独特的建筑世界。

HIROYUKI SUZUKI:
I see in your architecture things that contradict each other and yet making a very good harmony.
RICARDO LEGORRETA:
Well, I think we Mexicans are very lucky in this aspect because we have the combination of the Indian culture and the Spanish culture, and the Spanish, also with influence of the Islamic culture. So, these two cultures, in certain ways, have very strong contradictions, but you are able to learn and to find out that at the same time there are several things in which they coincide.
Like there, very strong forms, very large spaces. The wall and the color; they exist in both cultures. So I have found that the way that would help me design is to enjoy the contradictions in any aspect of life, because if you talk about-as I said before-about romanticists, and then you play with very big spaces, it seems that you cannot combine them. But it is this contradiction that help you to find the solution.
TADAO ANDO:
What do you think of the education of

就是去享受我们生活中各种彼此矛盾的因素。因为如果你谈到，正如我前面所说，关于浪漫主义者的问题，然后你又要面对一些很大的空间，这时候你似乎觉得无法将它们联系起来。但正是这种矛盾能够帮助你去找到解决问题的方法。

大学里教的，是对设计的强烈关心和有关实际建设的知识

安藤忠雄——我也看到过雷可瑞塔先生的几个作品。以前和美国建筑评论家肯尼斯·弗兰姆普敦❶先生谈话的时候，大家都对雷可瑞塔先生的设计给予了很高的评价。在全世界的建筑都变得非常相似的过程中，雷可瑞塔先生并不是将地域的传统和文化进行简单的引用，而是用另外一种形式进行重生，这一点非常可贵。

在雷可瑞塔先生大胆的造型和色彩中，有着与墨西哥风土不同的独自的世界。虽然大家经常谈到雷可瑞塔的建筑和路易斯·巴拉甘❷的建筑之间的关系，但我认为两者是完全不同的。我们能够从雷可瑞塔的建筑形体中感受到墨西哥人的热情，聆听到他们的述说，这一点让我深为感动。

在与雷可瑞塔先生的二十年的交往中，虽然直接

❶ 肯尼斯·弗兰姆普敦（Kenneth Frampton，1930— ）：英国建筑史学家，评论家，写过许多关于现代建筑运动的著作。他将在作品中积极反映地域特性的设计思想称为"批判的地域主义"。

❷ 路易斯·巴拉甘（Luis Barragan，1902—1988）：墨西哥具有代表性的建筑师之一。雷可瑞塔常被称为是巴拉甘德后继者，他们的建筑中确实有许多相似的东西，但是，两人并不是师徒关系，也没有直接的接触，雷可瑞塔本人也从未提及他受到过巴拉甘的影响。虽然两人的建筑在视觉上有相似之处，但也有许多不同。例如巴拉甘的建筑多局限于小住宅、教会等小型建筑，而雷可瑞塔则设计了加利福尼亚的豪宅以及巨大的郊外宾馆，这恐怕不能仅从时代背景的不同来解释。

architecture in Japan?
RICARDO LEGORRETA:
It is a very difficult question but one is very interesting how when you see the other side of the story you find some qualities.
One of the criticisms I do of the Western educational system is that there is the risk, not to educate architects but to educate designers. And you start to see on the Western countries that they only care about form and shape, and they don't know how to build. The extreme of that education, to me, is the United States, in which the architect is not anymore the head of the team. You wouldn't believe it, but for example, some very well-known architects of the United States don't go to the construction site.
So, I think we have to find an equilibrium of both things, because we are builders, we are constructors. But at the same, we are designers, we are architects. So we have to put together the two things: we have to know how to create spaces, how to handle light, how to handle proportions, etc. , but at the same time we have to know how to

交谈的机会不多,但通过互相赠送作品集等,我也学到了很多东西。从刚才的谈话中我感受最深的,是雷可瑞塔先生深深地热爱建筑,努力学习,并从经济和社会的因素中、地域性和传统中摆脱出来,获得自由,并从中开拓出新的世界。

美国加利福尼亚 Chiron 生命科学图书馆 (CHIRON LIFE & SCIENCE LABORATORIES Emeryville City, CA, USA) /1998
Photographer Lourdes Legorreta

因为今天来了许多学生,我想还是再问一些有关建筑教育的话题。一般来说,欧美的建筑教育,目标是培养建筑师,而在日本,不仅仅是东京大学,整个日本的大学都是在进行一种更加广义的建筑教育,其结果是在毕业之后,真正立志要成为建筑师,并决心为此奋斗终生的学生不多。我想雷可瑞塔先生对日本的建筑教育也有一些了解,能否请您谈一下对日本的建筑教育的看法呢?

理卡多·雷可瑞塔——这是个很难回答的问题,但当你换一个角度来看问题,会发现问题的另一些方面,这是很有意思的。

我对西方的教育体系的批评之一是这种教育存在着这样一种风险,即我们可能不是在培养一个建筑师,而是设计师。你仔细观察西方国家教育就会发现他们只关心形体和形状,而对建造的问题一无所知,我认为美国的教育就是这样的一种极端例子,那里的建筑师已经不再是一个团队的领头人。你可能不相

build them.

That's why I think it is so important that architects like Ando are coming to the university and that's why I was very happy and I'm very proud that you've invited me to be with you, more than come, give beautiful pictures and just give the lecture.

About 5 years ago, we changed completely the organization of my office. My son decided to be an architect, but he didn't want to work with me. So he finished his school and he came to Japan to work with Maki. Then he went to Spain and he worked Bohigas. And one day, I found the solution. I told him, "Bring all your friends to the office." There were all very young—22, 23 years old—and the people with experience in construction of the office stayed to teach them how to build. At this moment, my son and myself, we worked together in all of the projects. And the result is that the office works like a university, with that combination of the very strong interest for design and at the same time with the reality of knowing how to build.

Trying to answer Ando's question, sometimes

❶ 桢文彦（1928— ）：1952年东京大学毕业后留学美国，在可蓝布路可美术学院和哈佛大学修完硕士课程，1965年成立桢综合计划事务所。在执教华盛顿大学、哈佛大学之后，1979—1989年任东京大学教授。获得过日本建筑学会奖、普里茨奖、UIA金奖等奖项，是一位在世界上受到很高评价的建筑师。其主要作品有东京的代官山建筑群、幕张国际展览中心等。

❷ 玻海加(Oriol Bohigas, 1925—)，主要活动在西班牙巴塞罗那的建筑师。与 J. 马路德莱（Josep Martorell）、D. 马凯（David Mackay）一起成立MBM事务所。在弗朗哥政权时期，自称"巴塞罗那派"，提倡加泰罗尼亚的地域主义。

信，但是举个例子来说，有些非常有名的美国建筑师甚至从来都不去建筑工地。

因此，我认为应该在两者之间寻找一种平衡，我们是营造者、建筑者，但同时我们也是设计者，我们是建筑师。所以我们必须将两者联系起来：我们应该知道如何营造空间，如何控制光线，如何把握比例等等，而同时我们也必须知道怎样去建造它们。

这就是我非常看重安藤这样的建筑师进入大学参与教学的原因了，出于同样的原因，我对你能邀请我加入你的行列，而不仅仅是去展示一些美丽的图片、开设一些讲座，而感到高兴和骄傲。

已经成为建筑师的儿子说不愿意和父亲一起工作

理卡多·雷可瑞塔——大约在五年前，我们完全重组了我的事务所。我儿子决定做一名建筑师，但是他不想为我工作。所以他在完成学业之后，来到日本为桢文彦❶工作。之后他又去了西班牙为玻海加❷工作。有一天我想出了一个解决办法。我告诉他："把你所有的朋友都带到事务所来。"那都是一些二十二三岁的年轻人，然后由事务所中有实际工程经验的人教导他

雷可瑞塔父子
Ricardo Legorreta and his son Victor Legorreta, Lead designers of Legorreta + Legorreta
Photographer Alejandra Laviada

I think that old way of teaching of the master and the helper can work also. The problem is the number of people we're dealing with.
STUDENT:
I know there are lot of things that I want to do and I should do. At the same time I feel limits of my ability.
RICARDO LEGORRETA:
I had exactly the same question when I was young. It is very difficult, but at the same time the advice I can give you is that when you're young you have all the enthusiasm of the world. On the other hand, you have the kind of doubts or the need for direction. My advice is this : if you meditate carefully, you know what you want. Don't give yourself excuses.
Very often, we say in life, "I didn't have good luck"or "I need money so I have to work in another thing "or "my personal circumstances were very difficult ." But I can tell you—and Ando mentioned that I love architecture, and yes, I love architecture, and I can tell you that at this time of my life I still sometimes find myself asking those

美国新墨西哥州圣菲的住宅区
(ZOCALO RESIDENTIAL COMPOUND
Santa Fe, New Mexico) /2002

Photographer Lourdes Legorreta

们如何去真正的建造。同时，我和我儿子在所有的项目中都通力合作。这样做的结果使得工作室形成了一种类似于大学的气氛，对设计的强烈兴趣和对建造等现实问题的重视被结合起来。

就安藤所提出的问题来说，有时候我认为传统的、师傅带徒弟式的教育方式也是行之有效的。问题是我们所面临的受教育者的人数有多少。

我现在仍然有各种困惑，但是绝不会放弃建筑

学生——我提一个非常孩子气的问题，虽然我有很多想做的事情和必须要做的事情，但我感到自己的能力是有限的。

理卡多·雷可瑞塔——我在年轻的时候也有同样的问题。这是一个很困难的阶段，我可以告诉你的是，在年轻的时候你怀有对这个世界所有的热情。但另一方面，你也有各种各样的困惑，你想要知道方向。我的忠告是：如果你能很好地协调这一切，你就可以知道你所想了解的一切，不要给自己找理由。

一生中，我们经常这样说"我不够幸运"或"我需要钱，所以只能在其他事情上下功夫"，或者是"我

questions. So, don't give up your ideas. Our profession has a little bit of luck of being heroes, because if your objective in life is making money, it's much better to open an restaurant than being an architect. If you are here it's because you love architecture. So continue, even if the horizon looks black. Don't worry, the light will come one day.

You're very lucky because 'light' is here in this university, by having the possibility of the kind of teachers you have and the attitude of all of them. So, let's continue because we have the best profession in the world.

STUDENT:
Are you influenced by the traditional Mexicican way of construction?

RICARDO LEGORRETA:
Yes. One, for example, is the situation or the tradition of Mexico. We need to give work to people. So Mexicans love to build. So one of the very strong influences I have had is that I want to use as much labor and individual participation in the construction as possible.

的个人环境艰难"。但是我所能告诉你的是，安藤也提到过，这就是我对建筑的热爱，是的，我爱建筑，而且即使是现在，我有时候还会问自己这些问题。所以不要放弃自己的理想。

我们的专业使我们有幸拥有了成为英雄式人物的一点可能，因为如果你的人生目标是赚钱，那么与其成为建筑师，不如去开餐馆更合适。如果现在你们在这里，是因为你们热爱建筑，那么就坚持下去，即使地平线看上去那么黑暗，不要担心，光明总有一天会到来。

你们非常幸运，因为光明已经出现在这所大学中，由于你们拥有这样一些老师，以及他们所有人的见解。所以，让我们坚持下去，因为我们拥有这个世界上最好的职业。

西班牙毕尔巴鄂
SHERATON ABANDOIBARRA酒店室内设计（SHERATON ABANDOIBARRA HOTEL Bilbao, Spain）/2004

Photographer　Jaime Ardiles-Arce

墨西哥人是非常喜欢建造建筑的国民

学生——从今天的谈话中我们了解到雷可瑞塔先生在墨西哥的村庄中巡回，并受到村民们的色彩感觉和窗户的开启方式等方面的影响，那么您是否也同样受到了墨西哥传统建造方式的影响？

理卡多·雷可瑞塔——是的，比如说墨西哥的环境或

Another thing is that we Mexicans, because of our culture, we love solid things. We love stones, we love to hit a table and know that it's solid. So that makes you select certain materials and certain construction systems. The other thing, also, is that there are certain ways of living in Mexico that are very important, such as the indoor and outdoor life is very common in Mexico and that makes a very special way of building so you can achieve both things.

The other thing, also, that has influenced the way we build in the office is the concept of 'changing buildings.' I think that it's very difficult to design a building and certain types of building, and consider that there will never be changed.

TADAO ANDO:
Thank you so much.

RICARDO LEGORRETA:
I'm the one who have to thank you because I believe in young people, and I love to be with them. And when I come back because I will come back to Japan, and you don't invite me I will feel very sad. So I will let you know with enough time so we can

墨西哥蒙特雷商业管理学院
(BUSINESS ADMINISTRATION GRADUATE SCHOOL / EGADE Monterrey, Mexico) /2001

Photographer Lourdes Legorreta

者传统就是我受到的影响之一。我们必须将工作分配给别人。墨西哥人很喜欢修建房屋的工作。这种习俗带给我的强烈影响使我每次都尽可能多地投入人力，并且只要有可能，我就多多参加现场建造施工工作。另一种影响来源于我们的文化传统，我们墨西哥人都喜欢实体。我们喜欢石头，喜欢敲敲桌子以证实它们是个实体。这就导致了你去选择一种特定的材料和建造体系。

墨西哥人特定的生活方式是另一种很重要的影响，比如室内和室外的生活在墨西哥都是非常普遍的，这就会使我们设计一种特殊的建筑，从而满足两方面的需求。

此外，"不断改变的建筑"这一概念也影响到我们事务所的建造方式。我认为某种永远不变的建筑或者建筑类型是很难设计的。

安藤忠雄——非常感谢。

理卡多·雷可瑞塔——应该我感谢您才对，因为我相信年轻人的才能，我愿意和他们在一起。当我再次来日本的时候，我相信我一定会再来日本，如果你不邀请我的话，我会感到难过的。那时我就会有足够的时间，我们可以安排一下，我非常希望能看看这些学生的作品，并对它们作一些评价。谢谢。

organize it because I would love to see the work of these students and comment on it. So arigatou!

卡塔尔多哈的得克萨斯A&M工程学院 (TEXAS A&M ENGINEERING COLLEGE Doha, Qatar) /2006

1998年7月18日 （星期六）

雷姆·库哈斯
REM KOOLHAAS
（库哈斯没来）

在讲座的前一天，雷姆·库哈斯发起了40度的高烧，所以没能来日。我一直盼望着能够再次看到他那张严酷的脸，这次感到非常遗憾。不过，这种突然的变更也非常符合库哈斯的特点。

现在的学生都拥有优越的生活环境和温和的表情。我非常希望他们能够亲眼看到库哈斯那严酷的表情，哪怕只是一瞬，也会感受到对建筑的思索应该是这样一个苦苦寻求的过程、是这样辛苦的一件事情。

我们日本人已经养成了一个习惯，认为一切都会按照我们期待的那样顺利进行，特别是现在的学生，这种倾向更是明显，他的讲座的意外终止，如果把它看作是另一种声音在告诉我们："世界并不像我们想像的那么简单。" 怎么样？

借着这个机会，思考了许多关于他的事情，这对我和学生来说都应该是一个收获。我把这几天来关于库哈斯的思考，作为这个系列中期的一个小结来写一下。

雷姆·库哈斯是在实际进行建筑设计之前，由于1978年写作出版的《疯狂纽约》这本书而被世人所知的。

《疯狂纽约》中所写的理论与他实际建成的建筑不同，这些不同有时会令人感到困惑，但是，从建筑界的情况来讲，很少会有理论和实际设计都是一流的建筑师。而且，就他理论的独到之处而言，是试图在设计的过程中从根本上反思一贯的做法，并找出创造

库哈斯作品——美国IIT(Illinois Institute of Technology)学生活动中心　　摄影　王建国

性的方法,这样的内容非常多,因此,理论与实际的偏差就在所难免。不过,如果考虑他在现在建筑界的基础上所进行的飞跃的话,他或许更应该算作是认真地将理论与实践尽量靠近的建筑师。

　　库哈斯的建筑由于表现非常优美,所以吸引了很多人。但是,库哈斯的真正实力在于理论,这应该是众所公认的。他的理论在这二十年中,给了我们无法估量的影响,而且这个影响还将继续下去。这不仅仅是对那些随处可见的模仿者来说的,即使对那些和他形式完全不同的建筑师来讲,也是一个对自己的建筑表现进行反思的契机,给人带来紧张感。同时,他的工作也起到了另一个作用,就是探索建筑表现自由的极限。

　　拉·维莱特公园的设计竞赛方案虽然没有实现,

但与当年密斯的玻璃摩天大楼方案发表的时候一样，这个方案给我们带来了强烈的冲击。方案显示出了引导时代的形象与概念。

他做的法国国家图书馆竞赛方案，虽然只是让卵形和拧过的棒形等各种自由形状的物体立体地浮于玻璃圆管中，但看到它的时候却会有一种"未来"的印象。这并不是现代技术这条延长线上的未来，而是展示出建筑表现本身面向未来的可能性。从这个图书馆的双重结构中，我们也可以联想到他的理论和现实的双重构造。由于建筑艺术必须满足社会、功能、技术、经济的要求，所以与其他造型艺术相比相当滞后，他是在努力缩小建筑与其他造型艺术之间的差距。

《S.M.L.XL》这本具有绝对厚度的书，在内容上也向我们传递了绝对强烈的信息。这本书从建筑师一贯所写的领域中跳了出来，如果你用普通的方法去阅读，就会感到他从一开始就过多地堆砌了大量非常明了的东西。不过，与他的建筑相同，这些充满幽默的内容，就好像在告诉读者，现代社会是一个超大化的社会，我们根本不可能对此进行整体的把握和正确的对应，在这样的现实面前，只能放松精神参加进去，除此之外别无他法。

虽然库哈斯具有这么大的影响，但是他实际上只是在做自己想做的事，至少在表面上是这样。他根本没有肩负着整个建筑界的使命感，也没有对他的影响力负责的责任感，这让人感到他快活而又冷酷。

但是，库哈斯焦躁而又痛苦的相貌，以及每天都在试图缩小理论与现实的差距的奋斗，也向我们表露出他心中强烈的困惑。

（安藤忠雄）

1998年10月29日（星期四）10:30am—12:00am　主持：凯瑟琳·梵德雷

弗兰克·盖里
FRANK O. GEHRY

不管他的哪一个作品，一眼就能判断出是他的设计。他的设计就是这样，具有极强的独创性。虽然大家都了解他的建筑，但没有一个人去模仿。他在1989年获得普利茨奖，成为举世公认的建筑大师，但还没有一个追随者。这位孤高的建筑师，就是弗兰克·盖里。

从他的风格来看，人们会认为他是美国西海岸自由风格建筑师的代表，但实际上，他出生在白雪皑皑的北国——加拿大的多伦多（1929年）。随家迁往加利福尼亚的时候，他17岁。经历了南加利福尼亚大学、兵役、哈佛大学研究生院的学习，又先后在洛杉矶、巴黎的设计事务所里工作之后，他于1962年在圣·莫尼卡成立了自己的事务所。

盖里用胶合板、金属网、镀铅铁板等一些随处可见的廉价工业材料建成自己的作品，并以这种风格登场建筑界。那个被称为"廉价工匠"或"大海之家"系列作品的代表作，就是"盖里自宅"（圣·莫尼卡/1978年），那是一栋对房地产公司贩卖的庸俗住宅进行改造后建成的房子，他至今仍然住在那里。另外，他也把蛇和鱼等生物的自然形态直接搬入建筑设计之中，在神户就有他这样一个作品，名叫"鱼之舞"餐馆（神户/1986）。

20世纪80年代后期，从"维特拉美术馆"开始，他的设计开始转向奔放的形式，他用金属板建造一些交织在一起的复杂形体和歪曲的波形墙面，形成一个"铁的表皮"的新系列，并不断沿着这个方向拓展。1997年，作为这种设计的顶峰之作，他建成了"毕尔巴鄂·古根海姆美术馆"（1997年）。

（本江正茂）

迪斯尼音乐厅内部 © Gehry Partners, LLP　　迪斯尼音乐厅外部 © Gehry Partners, LLP

拥有自由的精神和无限的精力

安藤忠雄——最近，弗兰克·盖里先生设计的"毕尔巴鄂·古根海姆美术馆"❶受到世界的瞩目，我也认为这是一个意义深远的建筑。另外还有正在建设中的柏林会馆❷，虽然我们现在还只能看到一些外观，但我从它的外观中感到这栋建筑好像是把充满活力的"毕尔巴鄂"塞进了一个四方形的盒子里面。我想，这两栋建筑应该能够成为装点本世纪末建筑界的两颗明星。

大约二十年前，盖里先生的初期作品"盖里自宅"建成的时候我去看过。对于看惯了现代建筑的我来说，这栋建筑令我非常吃惊，这一点我记忆很深。当时那栋建筑给我的震动一直持续了二十年，这期间又一次次地看到盖里先生不断地向着新的目标挑战，我不断地观摩他的作品，感到受益匪浅。

对于盖里先生的建筑，不管从平面图还是立面图上，我们都很难理解他的形式和空间，圆柱形、圆锥形、立方体等各种形状扭曲、翻转，不断地碰撞组合在一起，产生一个不可思议的建筑。我想这种自由的设计思路，应该是从他在洛杉矶建造自己自宅的时候

❶ 毕尔巴鄂·古根海姆美术馆（1997），建于西班牙北部巴斯科地区的工业城市毕尔巴鄂，总展示面积超过11000平方米的巨大的美术馆，1997年秋开馆。主要收藏欧美20世纪的美术作品，特别是战后的美术作品。古根海姆财团在纽约建造了由赖特设计的古根海姆美术馆，"毕尔巴鄂"是其在欧洲的新馆。纽约古根海姆美术馆在开馆的当初被称为"混凝土桶"，而"毕尔巴鄂"则被称为"钛金属花瓶"。

❷ 柏林会馆（1996—2002年），指建于柏林的中心区，银行街巴黎士广场中心的"DG银行"。由于城市规划的严格限制，以石材为主的方形外观很难看出盖里的风格，但在玻璃覆盖的中厅里，放入了一个像狐狸头盖骨那样形状的会议厅。

KATHRYN FINDLAY:
When did you decide to become an architect and were there any formable influences in your middle and high school years? Can you describe how you feel that you have been affected by any of these particular experiences? Were there any teachers who had a significant impact on you and your choice of a career?

FRANK O. GEHRY:
I went to school till the end of high school in Toronto, Canada. And during that school I was interested in engineering—chemical engineering. I remember looking at the curriculum at the University of Toronto for architecture, and deciding that it was too boring.
The thing may be consider from that story is today Toronto tries to claim me as one of their 'sons' and they say architecture in Canada would be different if I had been there. And my answer is that it would not have been possible to become 'me' the way I am today had I stayed in Toronto. It's a question. I'm not sure, but it feels like the community does have a dampening effect on

就一直延续下来的,当时他发表了他的宣言:"建筑就是艺术。"

对于学习建筑或者相关专业的年轻人来讲,我们很想知道这种自由的思想的源泉在哪里,我们请盖里先生从这个话题开始,包括设计建造时的辛苦等等,给我们做一次讲话,我想我们一定会从中学到很多东西。

今天我们请凯瑟琳·梵德雷老师进行提问。几年前在架在泰晤士河上的桥❸的设计竞赛中,盖里先生和诺曼·福斯特先生,还有凯瑟琳·梵德雷女士一起进入了最后一轮评选,我想竞争对手之间进行谈话一定会有许多有趣的见解,所以今天就请凯瑟琳·梵德雷老师进行提问和对话。

凯瑟琳·梵德雷——今天很高兴能够请到弗兰克·盖里先生,我想他一定会有许多有意义的讲话。讲座的题目是"建筑与教育"。我想请他谈一下他独特的建筑形式是怎样形成的,他性格中的什么地方形成了他独特的建筑视觉。

从盖里先生的经历中,我们可以看出他是一个阅历和知识都很丰富的人。他出生在加拿大,最初是在南加利福尼亚大学学习美术,之后才开始学习建筑,并进入著名的维克多·格鲁恩❹事务所工作,然后又进

❸ 架在泰晤士河上的桥:指1997年千年桥(Millennium Bridge)设计竞赛。是连接泰晤士河北岸的圣·保罗大教堂前广场和南岸的格罗布剧场以及泰特美术馆的一座步行桥。福斯特在设计竞赛中获胜。该项目在2000年完成。

❹ 维克多·格鲁恩(Victor Gruen,1903—1980)。在奥地利出生的美国人。以设计地区性购物中心而著名的建筑师,城市规划师。

creativity by this very structure of its curricular and its attitudes.
I moved to California—and my family was very poor—so I was a truck driver for 2 or 3 years, and I went to school at night and I took some classes in fine art and met artists. And I took a class in Ceramics, and this is very important, the Ceramics teacher told me that I should look at architecture. And he was having a house designed by a famous Californian architect at that time, Raphael Soriano, and my teacher introduced me to Soriano. And it was like a light went on. I became interested. In my youth, when I was in high school I always was interested in painting and sculpture, and music-a classical music. My frist teachers in architecture in the Southern California were people who had been to Japan-GI's who occupied Japan after the war. And they came back as teachers, very much in love with the classical Japanese art and architecture. I studied Ise shrine before I studied the Parthennon; I studied Hiroshige before I studied Picasso. The climate and the character of Southern California lent itself to building with wood,

入哈佛大学研究生院学习城市规划,1962年成立了自己的事务所。可以说他一个人拥有了三个人的知识和阅历。

盖里先生精力极其充沛,我想一般的建筑师如果能够拥有他一半的精力就会满足了,可他还在说他还想做更多的事。在设计中,他经常打破通常的思路,不遵守固定的规则,是新事物的代言人。

我最近去了一趟英国,英国一般的报纸也对西班牙"毕尔巴鄂·古根海姆美术馆"评价很高,来参观的人数超过了预期。

盖里先生也有在各个大学执教的经历。他曾在耶鲁大学、哈佛大学、南加州大学等多所大学执教或讲演。特别令我感动的是,他曾经和研究生一起,就城市规划问题,向洛杉矶的高中生和学习环境的学生,以及参观美术馆的一般观众进行讲解。

今天,有着这些经历的弗兰克先生来给我们进行演讲,我们对他的谈话充满期待。

知 道 建 筑

梵德雷——您是什么时候决定作一名建筑师的,在您

and many of the early houses that inspired me were by Harwel Hamilton Harris, that name is no one here. And if you look at his work, you'll see that he was very inspired by Japan.

I also studied Japanese music, and I played in a gagaku orchestra at UCLA—I was the Clink. So I think from the beginning, in the American schools, because of my workbase, because of my early involvement in California, the Western classicism was not a high priority, and it already made me an 'outsider' to the architectural schools in general in America.

After I graduated in Southern California, I had to spend 2 years in the army, and then I went to Harvard University to study City Planning, not architecture. I had the feeling that architects could do more to change the world, in city planning. When you're young you have those notions. I hope you could too. It was at Harvard that I fell in love with Corbusier, and the reason was that they had a show in the studio of Corbusier's painting. And if you study those paintings, you realize that he was working out the vocabulary for

读中学期间,是否存在一些重要的影响力促使您做出这样的决定?您是否可以描述一下,当您受到这些特殊经历影响时候的感受?是否有老师曾经对您发生过重要的影响,并引导您选择了现在的职业?

盖里——我在加拿大多伦多市一直读到高中。在校期间我对工程学——化学工程产生了兴趣。我还记得当我看见多伦多大学建筑专业课程表时的情形,当时断定它非常的乏味。

也许是这件事,使得现在多伦多试图宣称我是他们的"儿子"之一,并且说如果我留在那里,加拿大建筑就会变得有所不同。而我的回答是如果我真的留在了多伦多,那么我就不可能成为今天的"我",这是一个疑问。我并不很肯定它的答案,但是教育机构所拟定的课程结构和他们的意见确实抑制了创造力的产生。

后来我搬到了加州,由于我的家境困窘,我做了两三年卡车司机的工作,我在晚上去学校,选修了一些艺术课程,并和一些艺术家相识。我选择的陶艺课对我起了非常关键的作用,陶艺老师对我说,你应该借鉴建筑。他自己拥有一幢由当时加州非常著名的一位建筑师拉菲尔·索里亚诺❶设计的住宅,我的老师把我介绍给了索里亚诺。这一切就像是打开了一盏明灯,我被深深地吸引了。

❶ 拉菲尔·索里亚诺(Raphael Soriano):建筑师,师从理查德·诺伊特拉(Richard neutra),被称为洛杉矶现代主义第二代建筑师,作品有"Case Study House"(1950),"Shulman House"(1949),"Colby Apartment"(1952)等。他利用端正的钢结构和大面积的玻璃形成典型的现代主义风格。美术青年盖里对建筑的兴趣竟然始于索里亚诺,这一点非常耐人寻味。

his architecture that led to Ronchamp. You can see it evolving in the painting. But as far as the rest of the school, the Planning School was boring; statistics, and government and politics and economics etc.

So I petitioned the school and became a 'free student' at Harvard. They gave me a card, and I could attend any classes, but I didn't have any credit. So I attended the architecture studios with Paul Rudolph and Edward Sekler who gave lectures on "the Golden Mean."I attended lectures very erwestly now that I quit the Planning School, I took many classes in government and politics. But at a nicer level, because there were government leaders who came to Harvard, and there were debates on philosophy-on Communism Vs Capitalism, Normon Thomas and other great speakers during the time.

After Harvard, I went to work for a Victor Gruen office in Los Angeles. It was a large commercial firm that had some idealism, and they were doing low-cost housing and city plans for Fort Worth, and the beginnings of the shopping center, which believe it or not those early shopping centers were very

在帕提农之前，先学习了伊势神宫

盖里——在我年轻时候，那时我还在读高中，我对绘画和雕塑产生了兴趣，此外还有音乐——古典音乐。我在南加州时候的第一批建筑方面的老师，曾经在第二次世界大战后作为驻扎在日本的美军士兵来过日本。回南加州以后，他们成了老师，并且都非常喜欢日本的古典艺术和建筑。我早在学习帕提农神庙之前就已经研究过伊势神宫了，我对歌川广重❶的学习也在对毕加索的学习之前。

南加州的气候和风俗使得当地人倾向于使用木材，那些曾经给我带来灵感的早期住宅很多都是由H.H.哈里斯❷设计的，这里不会有人知道他的名字。但是如果你看过他的作品，你就会知道他受到日本文化的很大启发。我也研究过日本的音乐，我在加利福尼亚大学洛杉矶分校的日本雅乐乐队中演奏铃铛。所以我一开始，即从美国读书时候起，由于我的工作背景，也由于我早期在加州的学习，西方的古典主义并没有对我产生一种主导性的影响，这使我成为美国大多数建筑学院的局外人。

❶ 歌川广重（1797—1858）：日本江户时代后期著名的浮士绘画家。其作品多为西方艺术评论家称道。梵高亦曾模仿其多幅作品。

❷ H.H.哈里斯（Harwel Hamilton Harris, 1926—1993）：建筑师。生于加利福尼亚，进行了大量的住宅设计，其设计既保留了加利福尼亚内陆木造住宅的传统，又拥有沟通内外的高大的流动空间。与前述的索里亚诺一样，虽然他曾在理查德·诺伊特拉事务所学习，但与其说他是现代主义，不如说与格林兄弟或赖特更有相似性。

idealistic ventures in trying to create a new town center-a new way of creating a town center in the growing suburbs of America. While I was working there, on weekends and evenings I also did small projects for myself; some houses and remodeling and learning to build actually, because the Gruen projects were so big-they were not accessible. While I was at the Gruen office as well as doing design I accepted responsibilities for management of teams and the financial management of the teams that I was responsible for. And while I was there I learned how to negotiate contracts with clients for big projects, and to negotiate fees and to organize the budgeting of work assignments so that the financial success of the projects in the office were assured. It was there I also learned to deal with the bureaucracies that governed the building process, and how to work with the consultants, the engineers-the structural, mechanical, acoustical, all engineers that collaborated. And one day, just as they were getting ready to offer me a very good partnership and permanent place with the office, I left.

在哈佛大学爱上了柯布西埃

盖里——当我从南加州毕业之后,我服了两年兵役,然后我就去了哈佛大学学习城市规划,而不是建筑。我当时认为建筑师在城市规划领域能更积极地改造世界。当你还年轻时,总是会有这样一些想法,我想你们也是这样。我在哈佛学习的阶段,爱上了勒·柯布西埃❸,原因是学校在工作室展出了柯布西埃的绘画。如果你曾经研究过这些绘画,你就会意识到他正在发明一种建筑语汇,而这最终创造了朗香教堂。你可以在他的绘画中找到它们的发展过程。但是除此以外,规划学院非常沉闷,课程包括统计学,以及其他关于政府、政治和经济的课程。

因此我向学校提出了申请,然后我就成为了哈佛的一个"自由"学生。他们给了我一张卡片,我可以参加所有课程,但是我不能得到任何学分。我和保罗·鲁道夫❹一起参加了建筑工作室,E.赛克勒为我们开设了几堂关于"黄金比例"❺的讲座。尽管我已经退出规划学院,我对这些讲座仍非常热衷,我还听了许多关于政府和政策的课程。都是些较高等级的课程,由当

❸ 勒·柯布西埃(1887—1965):活跃在20世纪20—60年代的建筑大师,现代建筑的鼻祖。对于建筑专业的学生来讲,与密斯、赖特一样,这是一个被首先记住的名字。他针对20世纪初盛行的建筑传统,以丰富的建筑形态发起了一场建筑革命。

❹ 保罗·鲁道夫(Paul Rudolph, 1918—):建筑师,生于肯塔基州。以现代主义为基调,将单纯的外观与丰富的内部空间进行组合,建造出具有雕刻感的建筑。作为教师,他广为人知的是任教于耶鲁大学。在哈佛大学教盖里的那段时间,正好是他将要去耶鲁大学之前。

❺ 黄金比例:在分割线段时,长段部分与短段部分的比等于全长与长段部分的比。$(\sqrt{5}+1):2 \approx 1.618:1$。从古希腊开始,就被认为是具有美感和调和的最佳比例,应用于美术和建筑之中。例如,米罗的维纳斯雕像,从额头到肚脐的长度与肚脐到脚尖的长度之比就符合黄金比例。

I went to Europe, and spent a year within Paris, and for the first time saw the great gothic cathedrals, visited Chantnes and traveled on weekends. I worked at an office in Paris, an architect's office, working in the French system-which is quite different from American. I spent weekends visiting all around the country, and became very interested in Romanesque architecture.
Having by then a strong interest in painting and sculpture of my contemporaries, I found in the Romanesque a way of integrating art and architecture in a very strong, tough way, not sweet-not like the baroque. It was very tough...it was very powerful. The work in Autun, the Romanesque church which the sculptor Ghizelbertus made the capitals. I just became very involved with not only the sculpture and how the structure was integrated into the architecture but also the painting of the Romanesque and how that was integrated into the architecture. And I spent 3 months, 4 months studying all those churches in South of France; Vézelay, Toulouse, Issoire, and more.
I went back to LA after a year in Europe,

时的一些政府官员来哈佛演讲，如诺曼·托马斯❶和其他一些伟大的演说家，同时也会开展关于社会主义和资本主义的哲学讨论。

> ❶ 诺曼·托马斯（Norman Thomas，1884—1968）：社会主义运动家，作为美国社会党主席，从1928年到1948年曾先后6次成为总统候选人。致力于失业保险、儿童劳动禁止、最低工资、缩短劳动时间等制度的法律化。

在建筑事务所里积累实际经验

盖里——结束哈佛的学习之后，我去了位于洛杉矶的维克多·格鲁恩事务所工作。这是一家大型的商业事务所，有些理想主义，他们当时正在为福特沃斯地区做一些廉价房和城市规划，以及早期的商业中心，不管你信不信，这些早期的商业中心是一些非常理想主义的冒险，它们试图成为新城的中心，而这在美国的郊区城市中心的建设中是一种全新的尝试。

在那里工作期间，我在周末和晚上也会为自己做一些小工程，包括一些住宅和改造项目，我试着学习一些真正的建造，因为格鲁恩事务所的工程项目规模非常庞大，不容易深入理解。在格鲁恩事务所的时候，我除了作设计之外也做一些管理工作，主要负责几个小组的管理工作和财政工作。在此期间，我学会了如何与业主谈判一些大工程，如何争取设计费，如何组织一个工程的预算，从而从经济上确保事务所工程的

and I started my own office. I had 20 dollars. I also had a family. You would laugh...you will laugh at my first works because they look like a pastiche of some Japanese temples. They are being published in the new book by Electa that comes out this month. If you get one, you can see the very first buildings I did in 1962, and you will have a lot of fun laughing at me.

At that time, Kahn was becoming very well known, and it was very hard to ignore him. I was also interested in the minimalism, in the work of Judd, Carl Andre, Smithson, and finally, Serra. And one of my first buildings, in the spirit of Kahn and minimalism, was a small studio called the Danziger Studio-that's also in the book. And it was on a very busy street in Los Angeles, and I made this very simple, calm building, like Ando, thinking that this would be peaceful in this chaos.

I later look back on that critically and felt that it was too separate from the environment that it was in, that it was; that I had ignored the context of the messy Los Angeles street. And I started to really look at Los Angeles-at the industrial buildings, the

❶ 巴黎的事务所：1961年，盖里32岁的时候在巴黎安德莱·鲁蒙迪（Andre Remondet）事务所工作。

❷ 位于欧坦（Autun）的罗马风教堂：指1146年建成的圣·拉撒大教堂。门上方的罗马风雕刻"最后的审判"是一个著名的作品。

❸ 威兹莱（Vezelay）、图卢兹（Toulouse）、伊索瓦（Issoire）：这些地方都有着12世纪建造的有代表性的罗马风教堂建筑。

❹ 自己的事务所：1962年，盖里33岁的时候开设了自己的事务所，名为"Frank O. Gehry and Associates, Inc."。

❺ 伊莱克塔（Electa）出版社的新书：Frank O. Gehry; The Complete Works, Francesco Dal Co, Kurt Forster, Hadley Soutter Arnold, feancesc Co, 1999

❻ 路易斯·康（Louis I. Kohn, 1901—1974）：建筑师。他的设计根植于现代主义风格，同时又继承了古典的美术学院的传统，创造出内涵丰富的静谧的空间设计，并以此著名。这里讲的20世纪60年代前半期的时候，他设计的"宾斯法尼亚大学里查斯研究所"、"索科生物学研究所研究楼"等建筑相继竣工了，并在孟加拉着手了一系列的项目。

顺利进行。在那里，我也学会了如何与掌管工程进度的政府部门打交道，如何与咨询师，以及参与工程的结构、机械、声学等所有工程师合作。直到有一天，他们正准备让我成为合伙人，并拥有事务所的永久职位的时候，我离开了事务所。

以艺术和建筑的有力结合为主导的法国时代

盖里——我去了欧洲，并在巴黎住了一年，我在那里第一次看到了宏伟的哥特大教堂，参观了那里的唱诗班，并利用周末旅行。我在巴黎的一个建筑事务所❶工作，它按照法国方式运作，与美国人的很不相同。我利用周末时间游遍了整个法国，并对罗马风建筑产生了很浓厚的兴趣。

我对当代的绘画和雕塑深深着迷，之后我又发现罗马风将艺术和建筑以一种非常强烈而明确的方式联系起来，两者之间的关系并不是甜美的，也与巴洛克艺术截然不同。它们非常的坚决，非常的有力。例如位于欧坦的罗马风教堂❷，雕塑家齐泽贝图为其设计了柱头。我不仅对雕塑有兴趣，同时也对建筑的结构如何融入建筑本身产生了兴趣，同时

圣·拉撒大教堂门上方的雕刻"最后的审判"
© S. Yamashita

也被罗马风的绘画，以及它们融入建筑的方式所吸引。我在法国南部花了3到4个月的时间去研究当地所有的教堂，这些地方包括威兹莱、图卢兹、伊索瓦❸等。

直面洛杉矶的现实，从中萌生出对自由的渴望

盖里——在欧洲住了一年之后，我回到了洛杉矶，并开设了自己的事务所❹。我当时只有20美元，还要养家。你可能会笑话……你多半会嘲笑我的第一个工程，它看上去像是对几个日本神庙混合物的模仿。它们将出现在这个月伊莱克塔出版社的新书❺中。如果你去买这本书，你就会看到在1962年我设计的第一个建筑，你将会从对我的嘲笑中得到不少乐趣。

当时路易斯·康❻开始变得非常出名，你很难忽略他。同时我也对极少主义❼产生了兴趣，比如贾德❽(Judd)、安德烈❾(Carl Andre)、史密森❿(Smithson)以及塞拉⓫(Serra)。在我早期的建筑作品中，有一个实验室的设计体现了康和极少主义精神，叫做但齐格(Danziger)实验室，书中也收录了。它位于洛杉矶一条非常繁忙的道路上，我设计了一座非常简朴、冷静的建筑，

❽ 贾德(Judd, 1928—1994)：典型的极少主义艺术家。他的典型作品就是用金属和混凝土建造的同样大小的立方体进行重复排列而成。

❾ 安德烈(Carl Andre, 1935—)：擅长将石头、木材、铁等材料在地板上进行同样大小的重复排列。与贾德彻底的几何形态相比，安德烈的作品更多地重视材料的变化（例如生锈）。

❿ 史密森(Smithson, 1939—1973)：建筑师。他在美国西南部的沙漠地区，探索适应野外环境的大规模项目，之后确立了大地作品(Earth work)、地景艺术(land art)的表现方法。

⓫ 塞拉(Serra, 1939—)：典型的作品是用铅棒或铁棒以及板等材料创造依靠重力保持平衡而给人紧张感的抽象雕塑。1970年以后，他的作品不仅出现在展览会上，也直接设置在城市空间之中。

streets, the mess of what was built and began to think of it as it is and tried to think of it optimistically instead of ignoring it, instead of denying it.

Having studied Frank Lloyd Wright and his use of the 30-60 module and Mies's use of modular construction, I felt that the grids that we would lay on the table under our drawings were too confining, and that I wanted something that was freer, more like the artists that I was friends with. So I forced myself to eliminate all of that and tried to draw and conceive of buildings in a different way. Now, the early buildings are still very rectilinear and they are just the beginnings-you have to really follow the progression to see when I get loose.

KATHRYN FINDLAY:
Actually, you've answered about 3 or 4 questions all at once. But I'd just like to take one step back and say, "What for you is architecture? "And this is definitely a leading question so I will explain.

In 1983, you were awarded a prize given to you, an architect who've made a significant contribution to architecture as an art. And

就像是安藤的，我想这会给嘈杂的环境带来宁静的气氛。

我后来以批判的眼光来看这一作品，我觉得它与它身处的周围环境太隔离了，它完全忽略了洛杉矶乱糟糟的街道形成的环境背景。然后我开始真正地观察洛杉矶这座城市，它的工业建筑、街道、以及混乱的建成环境，我开始思索，并试着以一种积极的态度代替过去对环境的忽视和否定。

在研究了赖特和他所运用的30～60的模数，以及密斯的模数之后，我感到工作台上的那些垫在我们草图下的格网太局限了，我希望能找到一种更自由，更接近我的那些艺术家朋友所使用的方法。所以我迫使自己排除所有这些方法，尝试用一种完全不同的方法来设计和思考建筑。至此虽然我的那些早期建筑仍然是非常直线的，但它们仅仅是开始，你将可以看到，放松束缚之后我所走的道路。

能够唤起观众情感的才是真正的"建筑"，哪怕这种情感是愤怒也好

梵德雷——事实上您一下子回答了三四个问题。但我想回过头来问一下，"您认为建筑是什么？"因为这确

your work demonstrates unique dexterity with form, a preoccupation for the shifting perspectives— and I think this is something that you're talking about— and at the same time it exerts a fascination with living forms, the snake, and the fish in particular. Is there a connection with your fine art training and your architecture? I think you're touching in this.

You have collaborated with a number of artists and creators. For instance, you collaborated with Richard Serra in a number of projects and you mentioned him. What are the benefits of such an interface? Do you think there is a clear boundary between art and architecture? Do you separate them? And is the 'fish' the leap-off point for your current work?

FRANK O. GEHRY:

Well, I always thought architecture was "to make buildings for people" and I like buildings that have 'feeling', that evoke feeling. I even like buildings that are 'angry', like Daniel Libeskind's new museum in Berlin, is very 'angry'. And you feel the 'anger' in the building. And that interests me because

实是一个最主要的问题，所以我想解释一下。

您在1983年被授予了一项大奖❶，而只有当一位建筑师做出巨大贡献将建筑推向艺术的高峰时，才会被授予此大奖。您的作品表明，您对形态的把握独一无二，那种视角变换的感觉非常强烈，我想这也就是您正在谈到的东西。同时它的魅力也让人联想到一些生物体，如蛇，尤其是鱼。您认为您所受到的艺术训练和您的建筑有关系吗？我想您正涉及到这些。

您和许多艺术家和创作家合作，如您的许多工程都是和R. 塞拉合作的，您也提到过他，这样的交叉合作会带来什么样的好处？您认为在艺术和建筑之间有着清晰的界限吗？您将它们分开对待吗？"鱼"是您近期作品的一个转折点吗？

盖里——我总是认为建筑是"为人建造房屋"，而我喜欢那些让人有"感觉"的房子，那些能被唤起的感觉。我甚至喜欢那些"愤怒的"建筑，像丹尼尔·里布斯金的柏林的新美术馆❷就是非常愤怒的，你可以在建筑中感到一种"怒火"。我之所以对这很感兴趣是因为，一幢建筑能让人们感觉到一些东西是很重要的，这也是我评价建筑的一个标准，即我是否能在那里感受到些什么。

对我来说，建筑是否是艺术并不是一个有趣的问

❶ 大奖：指的是美国AIA举办的布鲁诺纪念奖，盖里在1983年获奖。

❷ 丹尼尔·里布斯金（Daniel Libeskind）在柏林的新纪念馆：2000年开馆。

I think if we can make buildings that make people 'feel' something, then I think that's an important, and that's how I judge architecture, if I 'feel' something when I go there.
Whether architecture is art is not interesting to me. And the reason it's not is because if I'm with artists they call me an architect, and if I'm with architects they call me an artist. And in both cases it's a dismissive thing because for instance, Kenneth Frampton thinks I am an "artist manque." So these issues aren't terribly important about where the line is, because the models in history are there. Was Borromini an architect, artist? How would you decide? The little San Carlo alle Quattro Fontane church is a work of art as far as I'm concerned. It's incredible moment.
The artists I knew were dealing with form, and space, and context, and issues of social commentary relating to the time they were living in, except they didn't put windows in their art, or they didn't put toilets in their art. But, and they didn't have clients, but they had the similar kinds of pressures of

① 肯尼斯·弗兰姆普顿（K. Frampton, 1930—）：建筑评论家。从他的"批判的地域主义"观点来看，在世界各地都以同样风格建造建筑的盖里是应该受到批判的。参照90页注1。

② 波罗米尼（Franceseco Borromini, 1599—1667）：罗马全盛期巴洛克风格的代表性建筑师。这里讲的圣·卡罗小教堂（罗马，1638年开工）就是巴洛克建筑的代表作。它显示出动态空间的顶点，是一个划时代的作品。

题。这是因为当我和一些艺术家在一起时，他们称我为建筑师；而如果我与建筑师在一起，他们又称我为艺术家。在这两种情况下，都是无足轻重的问题，因为肯尼斯·弗兰姆普顿①就认为我是一个"不成功的艺术家"。这些例子说明界限并不重要，历史上就曾经有过这样的例子，你说波罗米尼②是艺术家还是建筑师？你怎么评判呢？我个人认为圣·卡罗小教堂是件艺术作品，它具有不可思议的意义。

就我所知，艺术家必须面对形体、空间和环境的问题，此外还有他们所处时代和社会对他们做出的评论，所不同的是他们不用在艺术品中涉及诸如怎么布置窗户和洗手间这样的问题。他们虽然没有业主，但是画廊、收藏家、美术馆也会给他们带来相似的压力，他们所承受的来自创造、遵从和超越的压力，和我承受的压力很相似。

被建筑界认为是怪人的我却得到了艺术家们的支持

盖里——在我的事业刚起步的那些年，当我需要支持的时候，帮助总是来自于那些艺术家而不是建筑师。我认识的那些建筑师总是觉得我很"滑稽"。而那些艺

galleries, and collectors, and museums, and the pressures to create, and to conform, and to excel, were similar to the pressures that I had.
In my early days, when you need support, the support for what I was doing came from the artists, not from the architects. The architects that I knew looked at me like I was 'funny'. The artists were always there, were always supportive, and I was very interested in what they were doing, how they were doing it, how they were accomplishing it, watching Richard Serra handle huge pieces of steel, watching Robert Smithson move earth, and Michael Heizer change a whole section of the American desert. These were powerful images and powerful processes too, that you couldn't ignore. But back in the architecture office, we had to still solve the issues of electric lights, plumbing, heating and air-conditioning, and relationship to the elements, and how to create buildings that the water doesn't come in, and so on. So there is a complexity that is added to out work that Richard Serra doesn't have to deal with. He has different issues to deal with. I

术家却总是支持我，同时我也对他们所从事的工作、他们的工作方式，以及他们如何完成作品很感兴趣。看着R.塞拉处理那些大块的铁片，或者R.史密森移动泥土，又或者看着米歇尔·海泽❸转换一片美国沙漠。这些都是一些非常有力的画面，也是充满力量的过程，不可能被人们所忽视。但是当我们回到建筑事务所，我们仍必须解决像照明、垂直、供暖、空调这样的问题，协调建筑内部各个元素之间的关系，以及如何让水不进入到建筑内部。所以在我们的工作中所加入的这些复杂性，是R.塞拉所无需面对的。他有他必须要解决的种种困难。我更多的提到塞拉是因为他是与我联系最紧密的一位艺术家，我们几乎每天都有联系，相互通电话、会面，探讨各种话题和作品，我们评论每个人的作品，除了我们自己的。我认为和艺术家的接触对我们建筑师来说是非常重要的。对我来说，这就等同于阅读一本书，保持与另一种感知力的接触，使自己就像是拥有了从不同角度观察形体和空间的神奇力量。

❸ 米歇尔·海泽（Michael Heizer，1944— ）：与史密森一起，同为大地作品的代表性人物。这里提到的是他从1969年到1971年之间设计的代表作——位于内华达州沙漠中的作品"双底片"，这是在一个山谷的两侧山峰上，分别挖出9米宽、15米深、30米长的壕沟。在此假想沙漠中存在着500米长的直线。

这样做的危险是你可能会陷入艺术家的领域，因为当你看着R.塞拉所创作的美丽铁片的时候，你会感到它是非常有诱惑力的。它看上去显得很简单、不费吹灰之力，因为它并不拥有一个建筑物所具有的复杂性，所以我经常会觉得这可能是一种更轻松的道路。

talk about Serra more because he's the artist who I'm most in contact with, on an almost daily basis: we speak on the phone and we see each other and talk about everything; everybody's work, we criticize everybody except ourselves. I look at the involvement with the artists as important for architects. For me it has been equivalent of reading a book, of having the constant discussion with another sensibility is like an amazing resource with somebody who's looking at form and space in a different way.

The danger is to fall into the artist's realm, because it's very seductive to watch Richard Serra make beautiful piece of steels. It seems very simple, very effortless, and it doesn't have all the complexity of a building, and so I often find that that would be an easy way too... But, you have to stay in your own place and realize there are many different forces at work in making a buiding.
KATHRYN FINDLAY:
You mentioned criticism and you mentioned in particular, criticism of unhelpfulness from fellow architects, and you mentioned

但是你必须坚守你自己的领域，并且意识到在一个建筑物生成的过程中存在着许多不同的力量。

正如每个人的签名都不一样，每个人的才能也不同，重要的是发现"自己的能力"

梵德雷——您提到了评论，还特别提及了来自建筑师的毫无帮助的评论，以及艺术家的支持带给您的鼓舞。在日本，我认为这是非常重要的问题，这里的建筑系学生受到建筑杂志和评论的很大影响。在您的职业生涯中，您不得不长期孤军奋战，然后人们才逐渐认识到您的成就的重要性。您能否谈谈自己是如何面对这些的，我想这是您曾经面对过的，也许还有其他的。您怎么看待建筑媒体？

盖里——在我还年轻的时候，我曾取消了所有送到我事务所的建筑杂志。我不会看这些东西，因为它们就像是一场时尚展示，你将会卷入一些与你所作所为无关的事务中，我想我无需它们就发现了这些。对于我来说，去发现自己的能力、自身所长才是更重要的，我对此非常渴望，我知道它们是与众不同的，但我不知道它们是否是优秀的，我甚至不知道自己是否会变得

encouragement in form of encouragement from artists. In Japan, I think this is particularly an important issue, but students of architecture are greatly influenced by architectural journalism and criticism. And in you career you have to battle alone for a long time and people recognize the significance of your achievement. Can you describe how you have dealt with I think you have, but may be some more. How do you think about the architectural media?
FRANK O. GEHRY:
Well, when I was younger, I cancelled all magazines that came to my office. I wouldn't look at them, because it's like a fashion parade and you become involved with something that's not of your doing, I think I found it without it. For me, it was more important for me, and I was very conscious of this; of trying to find my own abilities, my own capabilities. I knew they were different. I didn't know if they were any good. I didn't know if I was gonna be any good. I just knew that I had ideas that I wanted to pursue, and that if I look over my shoulder at what other people were doing I would get nervous. Ando

优秀，我所知道的只是我有想法，并且想要实现它们，如果我只是看着别人在做什么，我就会变得焦虑起来。我当时并不知道安藤，但是如果我在一本书上看到安藤做了一个更棒的建筑，我就会变得很沮丧，你明白了吧？

所以你不能让自己陷在里面。你必须塑造你自己的"空间"，因为我们每个人都是不同的，这就像是你的签名。当你写下你的名字的时候，你就能认出你的签名。我想我们每个人都有属于自己的性格，而这种性格能反映在你的建筑和作品中，我认为如果你对设计感兴趣，那么你一生中最重要的事情，就是去发现那些属于你自己的"语言"和"词汇"。你可能会发现你自己并不想成为一名设计师，你可能更想要成为一名技术方面的建筑师，你会发现在建筑领域，还有许多事情也是同样重要的，我的所有建筑都是由一个庞大的队伍来协作完成的，如果少了他们，我根本无法完成这些作品。在这个队伍中并不是每个人都有同样的天分。感谢上帝，使得他们才能各异，因为这样我们才能形成互补。所以我认为发现最适合你做的事情才是重要的，然后再形成自己的语言。我不认为你可以在传媒中发现这些。

wasn't there yet, if I looked at the book and saw Ando doing a better building I would get very upset, you see?
So you can't become involved with that. You have to forge your own 'space' because we're all very different, just like your signature. When you write your name, you can recognize your signature. I think that we have different make-ups, and those make-ups can be expressed in architecture, in your work, and I think the strongest thing you can do with your life is to find those 'your' language, your words if you're interested in designing. And you can even if you may find that you don't wanna be a designer-you may wanna be the technical architect, you may find other things in the field of architecture that are equally as important, because all of my buildings are done with a big team, a people who without them I couldn't do them. And they're not all of the same talents. They don't have the same and thank God they don't, because we can complsment each other. So I think the important thing is to find what's the best thing for yourself to do, and to find your own language. And I don't think you

盖里自宅外部
© Gehry Partners, LLP

盖里自宅内部
© Gehry Partners, LLP

捷克布拉格尼德兰大厦
© Gehry Partners, LLP

我在几年前已经停止购买阅读建筑杂志，也没感觉到信息不灵

盖里——当你年纪大一些，稳固一些的时候，你将能更好地欣赏同行的建筑作品。而当你还年轻的时候，很容易被影响，有的时候甚至会被这种影响淹没，有时候甚至会认为自己永远都不可能做得像他们那么好。这些就有可能会阻止你继续前进，因为要让我们所有人都能够按照自己的意愿去建造建筑，似乎是不可能的。

我也不建议人们转向过去，在某种程度上向过去的艺术家学习可能是更可取的方式，因为这样做带来的风险会小些。同样的，你也可以向其他领域学习，对我来说，向那些画家和雕塑家学习会更为安全，不会那么容易被影响。因为我不会想要，或不敢去做和他们相同的东西。

有一个试验值得你们尝试，就是停止读报一个月，在这个月结束的时候，你就会知道如何不通过报纸去了解所有重要的事情，这同样也适合于建筑出版物与建筑。我大约保持了4~5年的时间不去阅读杂志，虽然我不再花时间去看杂志，但还是可以知道那些重要事情。通过间接了解，你自然会得到那些信息。

find that in the media.
When you're older and more secure, you will appreciate your colleague's architecture much better. When you're young and impressionable, it can sometimes be overwhelming, and almost impossible that you will think if you'll ever be able to do something like that. So and it could prevent you form continuing, because it seems so impossible to make some of the building that all of us get to make.
Even though I don't suggest that one should turn to the past, in a way it's easier to look at the work of artists from the past, and somehow less threatening. Also, you can learn from other disciplines, and I suppose for me, looking into painters and sculptures was less threatening, was less overwhelming, because I never intended or pretended I was gonna do anything like it.
One experiment you can try, is if you stop reading the newspapers for one month, and at the end of the month you'll realize that somehow you know all the important things that happened then without having read a newspaper and it's the same in the architec-

我并不是让你们不去阅读和购买任何杂志,那样的话A+U❶一定会杀了我。我只是告诉你们,如何去发现自我,如何让自己获得自由。现在每个人都在以不同的方式发现自我,这正是我将自己从那些损害自身的重负中释放出来的一种方式。

❶ 日本出版发行的一本著名的建筑类杂志。
——译者注

想进行建造的冲动
和不知道会建成什么的恐惧
是设计和建造的原动力

盖里——在你们的文化中,建筑和传统有自己严密的体系,很细致,保持着自己的规范。然而书法和绘画却是很自由的。因此在你们的文化中……当然还包括景观艺术,有许多方面都可能激发你们去释放自我。我认为你们没有必要只按一种方式设计。这里有着各种选择。你可以欣赏一首诗,然后从中得到启发,你们也可以阅读普鲁斯特❷。到处都充满了灵感。你也可以走向一个具体的形体,然后产生一种要将这种体验转化为建筑的激情。

❷ 马歇尔·普鲁斯特(Marcel Proust, 1871—1922):法国小说家。作品有《挽回失去的时间》等。

艺术创作是一项依赖直觉的工程,你必须学着去相信自己的直觉。如果你对自己正准备着手的工作相当了解的时候,你就不会感到担忧。当我开始设计之

tural press and the architecture. Somehow, and I took no magazines for maybe 4 or 5 years, and somehow I knew the important events that were done without having to spend time with that. In your peripheral vision, somehow information comes to you. I'm not suggesting you don't read or buy magazines—A+U will kill me. I'm just giving you an idea for how to find your 'self' and how to find your freedom. Now, everybody's gonna have different ways, but it is one way that I've found to free oneself from the baggage that is traumatizing. In your culture, the architecture, the traditions, are very well organized and very careful and very prescribed. However, the calligraphy and the paintings are free. So there is in your culture many... and the landscape, there are many parts of that that are also possible to inspire you and to liberate you. I think you don't have to do it just one way. There is anything. You can read a poem and become inspired, you can read Proust. It's filled with inspiration. You can go to a concert and fad great impetus to make something as a result of the experience.

前，我总是充满了困惑。我不能肯定自己应该做些什么。你们可能也曾有过这样的经历，当你走进工作室，你所做的第一件事是清理你的工作室。你之所以这样做，是希望通过这种抵抗性机制去推迟富有创造性的工作，因为这让你感到慌张，让你感到担心。我认为这种情绪也是我们工作的一部分。之所以会产生这种担忧的情绪，是因为我们并不确定我们所要做的事。多年以后这种情绪已经能让我感到安慰，因为我知道如果清楚自己将会做些什么，我很有可能就会停滞不前了。

向弗兰克·盖里和安藤忠雄学习

梵德雷——即将成为建筑师的各位东京大学同学，下面我将作一下总结。在你们面前的是当今世界上少数最为出色的建筑师中的两位。他们对彼此差异性的交流和欣赏，证明了他们所具有的伟大而非凡的才能。他们都取得了自身的成功，安藤先生已经迎来了自己的时代。没有人会将他的作品与盖里先生的作品相混淆，而盖里先生也取得了同样的成功。他们各自在努力着，在自己的道路上奋斗。他们付出了巨大的努力

Making art is an intuitive process, and you have to learn to trust that intuition. If you knew ahead what you were going to do, you wouldn't be afraid. When I sit down to design, I'm always afraid. I'm not sure what I'm gonna do. You probably experienced going to your workplace and the first thing you do is you clean up the workplace. You clean it up because you use all kinds of denial mechanisms to delay starting the creative process because it's scary. You're afraid. And I think that's part of our work. There's this certain amount of fear because you don't know exactly where you're going. And I think that is comforting to me over the years, because I know that if I knew where I was going. I would stop, probably.

KATHRYN FINDLAY:
I would just like to say one thing as a comment is that, you know, Tokyo University students architect-to-be. In front of you are two men who are among the few most significant architects in the world at the moment. It's a testament to their greatness and their fantastic ability that they can share each other's differences, appreciate each other's

以形成自身的个性。没有人对他们提供过特殊的帮助。即使你们是东京大学的学生，也不能指望得到特殊眷顾，你们必须依靠自己。而且正因为你们是东京大学的学生，你们是有能力的，你们必须变得坚强起来。谢谢。

毕尔巴鄂·古根海姆美术馆／1997
© Gehry Partners, LLP

differences. They've all made their own way: Ando-san has had the time ready. No one recognized him as Gehry has also had. They both struggled; they fought for their own way. They've had enormous strength for character. Nobody had spoon-fed them. And you shouldn't look to be spoon-fed, your sense, even if you're from Tokyo University. And particularly because you're from Tokyo University, you've got the ability. Just be strong. Thank you.

1998年10月30日（星期五）2:30pm — 4:00pm　主持：岸田省吾

贝聿铭
I.M.PEI

如果提起在"卢浮宫美术馆"中建造了玻璃金字塔，和在"香港汇丰银行"（诺曼·福斯特设计）旁边设计了熠熠生辉的70层"中国银行大厦"的人，即使不知道贝聿铭这个名字的人，也会点头说："啊，我知道那个人。"

贝聿铭1917年生于中国广东。1935年转道日本赴美，后成为美国公民。在麻省理工学院、哈佛大学研究生院学习之后，受泽肯多夫之邀，进入韦伯 & 克纳普公司工作。1955年成立贝聿铭事务所，从1989年开始改名为贝－考布合伙事务所。

他从"国立大气研究中心"（科罗拉多，1967）的设计开始受到世人关注，之后又在"基督教科学中心"（波士顿，1973）、"华盛顿美术馆东馆"（华盛顿DC，1978）的设计中取得成功，成为著名的建筑师。在"卢浮宫美术馆扩建工程"（巴黎，1989）中，他在卢浮宫美术馆的中庭中建造了一个边长33米的玻璃金字塔，即使现在仍令到访的人感到惊愕。另外，还有"中国银行"（中国香港，1989）、"Miho美术馆"（滋贺县，1997）等作品。

在"华盛顿美术馆东馆"的建设中握有主导权的包尔·麦龙氏，看到贝聿铭设计的"国立大气研究中心"与周围的环境十分融洽地结合在一起，同时也被贝聿铭在工作中与科学家的配合精神所感动，因而指名贝聿铭进行"东馆"的设计；指挥"卢浮宫工程"的密特朗总统也是因为非常欣赏"华盛顿美术馆东馆"与现有建筑之间的调和关系，从而指名贝聿铭进行了"卢浮宫美术馆"的设计。

（冈由香利）

卢浮宫美术馆扩建工程／1989 © T.Bojo

华盛顿美术馆东馆／1978 © M.Chiba

汇集了现代技术精华的金字塔

安藤——今天贝聿铭先生的讲话也一定会让我们学到很多东西。我多次访问过贝先生设计的"卢浮宫美术馆",深受启发。卢浮宫一开始是作为城墙来建造的,后来成为王宫,18世纪末改为美术馆,经历了一次又一次的改造。我想,贝先生设计的这个金字塔,也将会在世界上最重要的文化中心法国的土地上留传几百年。卢浮宫在奥斯曼当年规划的轴线❶上,位于巴黎的正中心,在卢浮宫的正中心又建成了一个新的中心,我认为这是一件非常重大的事情。

当我第一次听到"在卢浮宫里要建造一座金字塔"这个消息时,我单纯地认为会是在卢浮宫的中庭建造一座混凝土金字塔,没想到建成的是这样一座玻璃金字塔,这让我非常吃惊。而且,这还不是一个简单的玻璃金字塔,它汇集了现代技术的精华,在它之后玻璃建筑在各地迅速发展起来,也能反映出它的意义确实非同一般。

我也在哈佛大学教过书。在波士顿市中心有一个"基督教科学中心"。这栋建筑是贝聿铭先生的代表作,

❶ 奥斯曼当年规划的轴线:从位于巴黎新中心的新凯旋门开始,在一条直线上能够看到旧凯旋门,香榭里舍大道、卢浮宫、协和广场,以及巴黎歌剧院等著名建筑。这条轴线是在拿破仑三世的时候,奥斯曼男爵从1853年开始推行了20年的大改革计划的成果。

SYOGO KISHIDA:
Today, I would like to ask several questions to Mr Pei about "architecture and architectural education ".
I. M. PEI:
Before I answer this particular question about architecture and architecural education, I think I would like to tell you something about where I am positioned in the modern history of architecture.
There may be a dispute as to when modern architecture really began. If you allow me to make a judgement on this subject, I would say it began with cubism. The reason cubism should be considered the origin of modern architecture as we know it is because it is a very broad movement. It permeates all branches of arts. painting, sculpture, literature, architecture. Cubism is a very important movement in the history of art. Therefore, I myself would prefer to say modern architecture did not begin with Frank Lloyd Wright. It did not begin with Mies van der Rohe. It begins with cubism. I think

它构思大胆，同时技术过硬。因此，虽然现在已经建成将近三十年，却没有一点老化的迹象。看到这栋建筑，我们这些从事建筑的人就会认识到，从建筑的耐用年数，包括空间的耐用年数和物质上的耐用年数上，能够反映出技术的重要性。

这一次在邀请贝先生给我们做这个讲座的时候，贝先生一开始并不愿意来，他说："我已经多年没有进行过讲演了，不去了。"但是我考虑到现在开始学习建筑的学生们一定会从贝先生的谈话中得到很多启示，就再一次请求并说服了贝先生，这一次贝先生告诉我，为了培养21世纪对社会有用的建筑师，如果时间不长的话可以考虑，他痛快地答应了我，今天来到我们这里。请大家一定认真听取贝先生的讲话，不仅要学习他作为一名建筑师的才能，更要学习他经过长期的学习历练，抱着顽强的信念生存下来的精神。

现代建筑开始于立体主义

岸田————在讲演之前，我想先简单地介绍一下贝先生的经历。贝先生生于1917年，今年81岁，生于中国。18岁时来到美国，最初在宾夕法尼亚大学学习，后进

it is important. It would not be agreed upon by all, but I would like to take that position first. I am sure you all know about Picasso, about Braque, or Juan Gris. Those are the men that opened the door for architecture to enter. Beaux-art school architecture which I think we no longer have anything to do with, I am sure you know. Nevertheless, they are powerful until cubism came along.

But that time, you have, particularly in Europe, a number of places where a change began. You have Tatrin in Russia, using simple example as a man, you have Rietveld in the Netherland, Utrecht, and you have Gropius in Germany, Mies van der Rohe in Germany, and you have Le Corbusier in France. These are the people that took up the call from cubism. And we say now this is a moment of change.

That is an introduction of my belief about the modern history of architecture: I would like now to say something to you about where I belong. I belong to the generation which follows Le Corbusier, Gropius, Mies van der

入麻省理工学院，然后又转到哈佛大学，最后在哈佛大学拿到了硕士学位。在这之后的成就我想不用讲大家都很清楚了，现在他以纽约为主要所在地进行设计活动。从世界范围来讲，贝先生是与丹下健三❶老师同时代的人，是紧接着勒·柯布西埃和格罗皮乌斯这些现代建筑先驱之后的一代人。

今天我将就"建筑与建筑教育"这个主题问贝先生几个问题。

贝聿铭—— 在回答关于建筑和建筑教育这个特定问题之前，我想先谈谈我在现代建筑史中所处的位置。

关于现代建筑真正的起始时间这个问题，现在还存在着争议。如果由我来判断，我认为它开始于立体主义❷。我之所以认为立体主义是现代建筑的开端，是因为正如大家所知，立体主义是一场非常广泛的运动，它渗透到各种艺术形式中，如绘画、雕塑、文学和建筑。立体主义是现代艺术史上非常重要的一场运动，所以我个人倾向于认为，现代建筑并非始于赖特，也不是始于密斯，而是始于立体主义。我认为这一点是非常重要的。尽管并不是所有人都这样认为，但是我想先阐明我自己的立场。你们肯定都知道毕加索，知道布拉克❸，或者格里斯❹。正是他们为建筑开启了大门。我想你们也知道巴黎艺术学院的建筑教育，我认

❶ 丹下健三（1913—2005）：可能是世界上最著名的日本建筑师。"广岛和平中心"（1956）的设计确立了他在世界上的地位。借日本高速增长期之势，先后设计了"旧东京都厅"（1957）、"代代木体育场"（1964）等公共建筑，近期又建成了"新东京都厅"（1991）、"新宿公园大厦"（1994）等建筑。

❷ 立体主义：为了更加清晰地表达作品，将对象进行分解重组，并不是表达瞬间的形象，而是表达自己的主观意志。另外，也把绘画作为一种平面化的形态艺术，是一场艺术运动。由毕加索、布拉克等人发起的这场运动后来被未来派继承，也是以蒙特里安为首的风格派诞生的基础。

❸ 布拉克（Georges Braque, 1882—1963）：画家。受到毕加索的影响，同时也受到赛尚将色彩分解成小斑点的做法的启发，是立体主义抽象绘画的旗手。代表作有"巨大的裸体"（1908）等。

❹ 格里斯（Juan Gris, 1887—1927）：画家。住在毕加索工作室的隔壁，受到立体主义的影响。

Rohe. We learn. We also from time to time disagree as you do today.
I remember when I was just a student at Harvard under Gropius, Gropius told us that the international style is the way to go. A globaliation of architecture was predicted by him. That would happen. One day he said to me, "you will find that modern buildings design in Europe and America will be followed by other countries, such as Japan". I protested it. I disagreed at that point. I was a student. Because I came from China, I would like to think China some day would not be building buildings like they would do in the United States anywhere, and in Europe. But I was wrong. Look what happened to Shanghai. Look what happened to Tokyo. Look what happened to Osaka. Gropius was right. (laughter)
But I have not given up. I continue to believe the world is much more interesting than to be globalized in that manner. First of all, to build a building you have to have a site. And the site has to be in a

131

为它和我们已经没有任何关系了。尽管直到立体主义出现之前，它的影响仍然是非常强大的。

在那个时代，许多地方都开始发生变化，尤其是在欧洲。我们可以简单地以一些人来说明，在俄国出现了构成主义，在荷兰的乌德勒支有里特维尔德❶，在德国有格罗皮乌斯和密斯，在法国有柯布西埃。这些人正是受到了立体主义的启发。我们现在会说，那是一个变化的时代。

❶ 里特维尔德（Gerrit Thomas Rietveld，1888—1964），以荷兰的乌德勒支地区为中心，活跃在建筑、室内设计等领域的建筑师，是以彼德·蒙特里安为首的风格派运动的主要成员。主要作品有"斯莱德邸"（1924）、"红蓝椅"（1918）等。

抗议格罗皮乌斯关于国际式风格将会全球化的预言

贝聿铭——这就是我对现代建筑史的理解，现在我将告诉你们我所处的位置。我属于跟随柯布西埃、格罗皮乌斯和密斯的一代人。我们向他们学习，我们所做的事情和今天你们的所作所为经常是不一样的。

记得在哈佛学习的时候，那时我还是格罗皮乌斯的学生，格罗皮乌斯告诉我们国际式就是我们应该追随的风格。他预言了一种全球化的建筑样式。有一天他对我说："其他国家将会追随那些出现在欧洲和美国的建筑设计，如日本。"我无法同意他的意见。我不能同意这样的观点。我当时是一个学生，我来自中国，

place. And the place has a history. And that history is extremely important to architecture and to architects. My definition of history is not the same as the defintion made by the post modernists. It is a different kind of definition. I want to say that. The post modernists, and I think you know what I mean by that, the movement took place in the 80s. You all must be familier with it. I think they were picking more the visual manifestations fo history, not really the big roots of history, which is what interests me. I think that is introduction, a long one. I think I have said enough. Now, I am prepared to discuss architecture with you.

SYOGO KISHIDA:

Did you have any reason or motivation to become an architect?

I. M. PEI:

I hope I do not disappoint you by saying that I really did not know when I left China, what architecture was. I thought architecture involves building. And it was only after I went

我不希望那些在美国和欧洲到处可见的建筑也出现在中国。但是我错了，你们看看在上海、东京和大阪所发生的一切，就可以知道，格罗皮乌斯是对的。（笑）

但是我并没有放弃，我仍旧认为世界要比现在全球化的局面要有意思得多。首先，要建造一座建筑就必须有场地，而这个场地必须在一个地方，而一个地方必定有自己的历史，而这个历史对于建筑和建筑师来说是非常重要的。我对于历史的定义和那些后现代主义者是不一样的，我们的定义截然不同。那些后现代主义者们，我想你们都知道我所指的是发生在（20世纪）80年代的那场运动，你们对此一定也很熟悉，我认为他们更多的是从视觉上来阐释历史，而没有深深地根植于历史之中，而后者才是我感兴趣的内容。这就是我的自我介绍，一个很长的介绍。我想我已经说得够多的了。现在我想和你们一起讨论建筑。

遇到伟大的教师格罗皮乌斯

岸田——有什么特殊的原因或动机促使您成为一名建筑师吗？

贝聿铭——当我离开中国的时候，我对建筑还一无所

to America that I discovered there are many disciplines which are involved in building. You can be an architect. You can be an engineer. You can be a builder. I had to choose.
Now I disappoint you. I choose engineering, architecture engineering, and not design (laughter).
SYOGO KISHIDA:
Fascinating! I heard you first studied at the University of Pennsylvania, being disappointed of Beaux-art methods, then moved to MIT in order to become an engineer. Yet you were not satisfied of MIT. Again you moved to Harvard because Gropius had seemed to be doing something stimulating. I suppose, meeting Gropius was a great impact to you.
I. M. PEI:
Yes, professor. I was disappointed in Beaux-arts methods of teaching architecture. As soon as I heard Gropius was coming to Harvard, I thought there is a possibility for me to learn something different from beaux-

贝聿铭先生在苏州河上
Photographer 林兵(Lin Bin)

知,我希望这样的回答不会让你失望。我觉得建筑和房子有关。我是在到了美国之后才发现,这里面还有那么多的门类。你可以成为一个建筑师,也可以成为一个工程师,还可以成为一个建造者。我必须对此加以选择。

现在你可能感觉失望了,我选择做一名工程师,建筑工程师,而并非建筑设计师。(笑)

岸田——很有意思!我听说您一开始是在宾夕法尼亚大学学习,您对那里巴黎艺术学院式的教育感觉失望,然后才转到麻省理工学院去学习建筑工程。但是您对麻省也不满意,所以又进入哈佛大学,因为当时格罗皮乌斯似乎在做一些让人兴奋的事。我猜,遇见格罗皮乌斯对您产生了重大的影响。

贝聿铭——是的,教授。我对巴黎艺术学院的那套教授建筑的方式感到失望。当听说格罗皮乌斯来到哈佛,我想这可能是和巴黎艺术学院式教育不同的一种学习途径。后来在哈佛的学习确实没有让我失望。我必须说我从格罗皮乌斯那里学到的东西要远远多于在麻省。不过格罗皮乌斯并不是一位伟大的建筑师。(笑)但是他的确是一位伟大的老师。

包豪斯的教育方式❶是非常宽泛的,而这种广泛性也使其能够成为一种很有效的教育方式,尤其是在早期。

❶ 包豪斯的教育方式:1919年创立的包豪斯采用了"将所有的艺术创造、工作技术、工学训练等统合成为建筑的基础"这样一种新的教育体系。参见第78页注3。

arts methods of teaching. And I was not disappointed. I did go to Harvard. And I must say that I learnt much more from Gropius that I did from M.I.T. But, Gropius is not a great architect.(laughter) But, he was a great teacher.

The Bauhaus method of education is a very broad one. And this is what make them so powerful, particularly during those early years. It included in the school painters. Many of them became very very famous like Paul Klee, like Schlemmer, like Moholy-Nagy. It included furniture designers, the highest craft tradition, as you have in Japan, People like Breuer. His furniture in the early years was as successful as anything he ever designed. In other words, it is a multi-discipline school. And it is gathered together. All disciplines important to architecture itself.

So, I think we have to look at Bauhaus in that way. It is not individual. I would like to cite Le Corbusier. It is called individual. It is a collection of disciplines, collection of

❶ 保尔·克利(Paul Klee, 1879—1940),从1920年到1931年在包豪斯学校教授彩绘玻璃、纺织、绘画和基础课程。

❷ 施莱默(Oskar Schlemmer, 1888—1943),从1920年到1929年在包豪斯学校教授石雕、金属造型、舞台造型等课程。同时也在芭蕾舞中探索新的表现方法。

❸ 莫霍里·纳吉(Laszlo Moholy-Nagy 1895—1946),从1923年到1928年应格罗皮乌斯之邀,在包豪斯学校教授金属造型、活跃在舞台装饰、广告设计等领域。与格罗皮乌斯共同编著《包豪斯丛书》。

❹ 布劳耶尔(Marcel Breuer, 1902—1981),生于匈牙利。1920年进入包豪斯学校,成为第一届学生,1928年离开包豪斯学校,其间主要活跃在家具设计领域。在柏林,伦敦进行设计工作,1937年在格罗皮乌斯的召唤之下,转移到哈佛大学,指导了约翰逊(Philip Johnson)和鲁道夫(Paul Rudolph)等人。主要作品有"联合国教科文总部"(巴黎,1958)、惠特尼美术馆(纽约,1966)等。

学校里有教授绘画的老师,包括保尔·克利❶、施莱默❷和莫霍里·纳吉❸;还包括了家具设计师,他们具有像你们日本人那样高超的传统手工艺,如布劳耶尔❹。布劳耶尔早期所设计的那些家具和他所设计的任何东西一样都很成功。换句话说,这是一个拥有多种学科门类的学校,它们汇聚在一起,所有这些对于建筑本身来说都是非常重要的。

所以我认为应该这样看待包豪斯,它并不是一个孤立的个体。我可以用柯布西埃式的教育来说明,它虽然被看成一个独立的个体,但是它是多种学科的汇总,聚集了各种类型的专家,他们以同样的方式,向着共同的目标前进。

岸田——您提到格罗皮乌斯不是一位伟大的建筑师,而是一位伟大的教育者。您是否仍旧认为自己深深地受到作为一位老师的格罗皮乌斯的影响?

贝聿铭——是的,我确实这样认为。他教会了我如何去分析,如何在开始工作前去分析问题。我认为这种方法勿庸置疑是非常重要的。但是在接受包豪斯教育的早期,我并没有接受这样的教育,而是被教导去临摹一种设计模式。但是格罗皮乌斯却没有设计模式:也许水平屋顶算是一种模式,但是却不存在一种设计模式。你必须去分析,你必须去证明你为什么这样设

specialists that work in the same way and work to the same goal.

SYOGO KISHIDA:

You mentioned that Gropius was not a great architect but a great tercher. Do you still think you have largely influenced by Gropius as a tercher?

I. M. PEI:

Yes, I would say so. I think he taught me to make an analysis. How to analyze the problem before you begin. I think that approach is obvious. Obviously important. But in the early years of Bauhaus training, I was not taught that. I was taught to follow a design model. But Gropius does no design model: perhaps flat roof, but no design model. And you have to make analysis, you have to prove to him why you designed that way. And that, I think, is simple, but it is very important to me.

Frank Lloyd Wright and Le Corbusier were great architects. They were exceptional architects in architecture history in my opinion. But they were not teachers. That I can tell

计。我认为这虽然非常简单，但对我却非常重要。

赖特和柯布西埃都是伟大的建筑师。在我看来，他们都是建筑史上非常突出的建筑师，但他们却不是教育者。我有理由这样说，因为他们两个我都认识。

年轻的时候想做实际工程设计

岸田——贝先生，您出生在中国，然后移民到美国，又在欧洲、美国和世界各地包括日本进行建筑设计。您的设计为世界各个地方所接受，并广受欢迎。您是否认为这与您的中国文化背景有着某种联系？

贝聿铭——在我回答这一特定的问题之前，我想先谈一下我作为建筑师的一段往事。第二次世界大战后，美国的建筑设计工作量非常有限，但是对住宅的需求量却很大。我为一位名叫泽肯多夫的房地产商工作，你们中的有些人可能对他有所了解。我和他合作了12年，从1948年到1960年。我和这位房地产开发商的合作非常紧密。

美国建筑师协会（AIA）往往会瞧不起与房地产开发商的合作。事实上他们并不主张建筑师这样做，但是我却这样做了。我想指出的是，在若干年大学的训练中，我从格罗皮乌斯和其他一些人那里受益良多，

you, because I knew them both.
SYOGO KISHIDA:
Mr. Pei, you were born in China, moved to the United States, designed architecture in Europe, US and all over the world including Japan. Your design is accepted and welcomed to the world. Do you find any relation to it with your Chinese cultural background?
I. M. PEI:
Before answering this specific question, I would like to say something first about a piece of my history as an architect. After the Second World War, there was very little work in the United States, but there was a great need for housing. I went to work for a real estate developer whose name is William Zechendorf, when some of you may have known of. And I associated with him for 12 years, from 1948 to 1960. I worked very closely with this real estate developer.
To associate with a real estate developer is something that the American Institute of Architects looked down upon. In fact, they discouraged architects to do that, but I did

除此之外和房地产开发商的合作也是另外一个使我获益匪浅的途径。

我最大的缺憾是没有机会去做设计，1960年的时候我已经43岁了，而我在当时却没有机会去做设计。为什么我今天仍在做一些本该二十年前完成的事情，我想这也是原因之一。我之所以这样说，是为了让那些追随我、安藤教授、理查德·迈耶❶以及弗兰克·盖里的学生知道，他们开始设计的时间要比我早得多，他们比我幸运得多，能比我早得多地发现自我。

❶ 理查德·迈耶（Richard Meier，1934— ）：从年轻时到现在一直在不断地建造白色的建筑。顺便提一下，他1965年在康奈迪克州设计史密斯住宅的时候只有30岁，十分年轻。

在剩下不多的人生中，我慎重地选择工作

贝聿铭——从1960年到1989年，这该是多少年了，将近30年，我有了实践的机会，我建造了不少房子。不错，的确有相当多的建筑，但是据我自己的理解，它们算不上是真正让我感到骄傲的建筑。因为我当时所接触的都是一些规模很大的工程实践。当有200或300人在为你工作的时候，你就不可能将注意力完全放在设计上，因为你必须将主要的注意力放在如何支付这两三百人的工资上。（笑）

我想这也是实际工程中你们应该了解的一个问题。

it. I want to mention that, because aside from a university training, and I gave you already my sense of debt to, Gropius and others, this was second.

For my biggest lack was that I had no opportunity to do design. 1960, I was already 43 years old, and I did not have an opportunity to design at that time. I think that is one of the reasons why today I am doing what I should have done 20 years ago. I say that, because I think it is important to know that the next generation that followed us, Prof. Ando, Richard Meier, even Frank O. Gehry, they started design much earlier. They had an opportunity to find themselves long before I was able to do so.

From 1960 to 1989, how many years is that, almost 30 years, I have practiced and I have built buildings. Yes. Quite a few buildings, but these are not really, in my opinion, buildings that I consider with pride. The reason is that I had at that time a very large practice. And when you have a practice 200 or 300 people working for you, it is very difficult to

你希望获得成功，你希望去完成一些大型工程。但是要完成大工程，你就必须有大事务所。如果拥有了大型事务所，你就会成为它的奴隶。而在1989年之前，我就是大型事务所的奴隶。（笑）

现在，我肯定就要回答到你的问题了。的确，我在世界各地都有许多工程项目，这是一些小型项目，并不很大，这些都是我自己选择的项目。在1989年，我从自己的事务所中退休了。由于我的工作开始得较晚，我本该可以更早地做出这样的决定。而现在我正在这样做。我只有两个助手。我没有大办公室，我不能拥有它。（笑）只能有两个助手。但是这却使我能够自由地进行选择。在需要的时候，我也可以请人帮忙，但是我必须支付一个非常非常高的价格。我宁愿这样。因此从1989年，或者比这更早的五年前开始，我决定自己选择业主，我的业主，这是非常重要的，甚至比工程本身还重要。我认为自行选择业主，这很重要。

当然卢浮宫的项目并不属于这一类，我并没有选择自己的业主，而是我的业主选择了我。而华盛顿的国家美术馆设计是在1968年完成的，那是30年前的事了，我也没有自己选择业主，他们选择了我。我对这些作品是比较满意的，它们开启了我生命中的幸运之门。但是

concentrate on design, because the major concentration in such a practice is how to pay 200-300 people. (laughter)

That, I think, is the kind of problem in a practice you should know. You want to be successful, you want to do big work. But to do big work, you have to have a big office. To have a big office, you become a slave to the big office. And I was a slave until 1989. (laughter)

Now I am sure, I am answering your question. Yes, I am doing a lot of work all over the world, small projects, not big ones. It is all my choice. I retired from my firm in 1989. And I decided since I started late, I am making up for what I should have done earlier in life. And I am now beginning to do just that. I have only two assistants. I do not have big offices, I could not(laughter). Only two assistants. But that frees me to choose. If I need help, I can get it. But I have to pay a very very high price. I prefer that. So, therefore, I decided then in 1989, or actually 5 years before then, that I would

事情并非总是如此。不过最近这些年中,不走运的日子变少了。我现在总是非常仔细地挑选自己的工程。

现在,你们可以向我提问了。

在18岁时看到的最初的"世界",是1935年的日本

学生——贝先生,您是在18岁的时候离开中国去美国的。当您看到新世界时,您有什么感觉?

贝聿铭——我必须承认,当我能离开家乡,去见识整个世界的时候,感到非常兴奋。1935年,我先看到了日本。我当时在一艘船上,行程开始于下关,让我想一下,是下关吗?是的。我在那里停靠,而船继续向横滨进发,我没有留在船上,而是搭乘了一辆火车。我访问了神户、京都和东京。然后我到了横滨。因此我在看到美国之前,先看到了日本。

1935年在东京的时候,我住在赖特设计的帝国饭店❶里。对,就是这样。在我置身于这个世界的时候,让我感到兴奋的是我将要遇到的各种冒险。这里确实存在一些风险,因为你们知道,日本政府当时可以阻止我的旅程。你们不要忘了那是在1935年。(冈田新一先生在听众席中笑了❷。)我猜你们不会忘记,因此

❶ 帝国饭店(1922):赖特设计的建于东京日比谷的宾馆。开馆的第二天就碰到了东京大地震。1967年,为了土地的有效利用,该宾馆被拆除,建筑的一部分搬到爱知县明治村保存下来。在植木等人主演的"没有责任心的日本男人"(1962,东宝)这部影片中有在帝国饭店里摄影的镜头,值得一看。

❷ 冈田新一先生在听众席中笑了:贝聿铭先生注意到听众中的一个人就是设计"最高裁判所"、"东大医院外来医疗栋"的建筑师。冈田先生在讲座开始的30分钟之前来到会场,坐在学生当中。"虽然以前和贝聿铭先生没有见过面,但透过他的作品可以看出他的人品,例如从他的雕塑性空间的制作方法中"(冈田先生讲)。

choose my clients. My clients. That is more important than the projects. I want to choose my clients. That is important.

Of course the Louvre does not fall into this category. I did not choose my clients. My clients chose me. Also the National Gallery in Washington was designed in 1968, 30 years ago. I did not choose my client. They chose me. so, therefore, I have had some satisfaction in work, which happened to be my good luck. But, this was not always the case. But, now with a number of years this lack of luck is shrinking. I am much more careful now in selecting my work.

Now, ask me questions.

STUDENT:

Mr. Pei, you moved to the US from China at the age of 18. What did you have in your mind before seeing the new world?

I. M. PEI:

I have to say it was a moment of great excitement for me to be able to leave my country, to see the world. First of all, I saw Japan in 1935. I took a boat. I started at Shimonoseki,

这在当时是一件相当冒险的事。我们中的两个人决定了我们下面的行动，我们并没有从船上跳下来，而是请求船长给我们4到5天的时间，然后我们和他们在横滨会合。由于船长并不鼓励我们这样做，所以我们完全是自担风险。你们瞧，即使是这样的风险也难以阻止我们想看看新世界的急迫念头。然后我们又经过了夏威夷、旧金山，接着就是新世界，它们是我生命中非常珍贵的一刻，我至今仍然记忆犹新。

与包豪斯相比，赖特在空间认识方面更加接近立体主义

学生—— 您刚才说现代建筑开始于立体主义。您认为立体主义是怎样影响包豪斯和柯布西埃的？

贝聿铭—— 包豪斯所理解的立体主义和柯布西埃的立体主义是有所不同的。我想这样说，立体主义的空间在柯布西埃那里得到了很大发展，这就是格罗皮乌斯，甚至是密斯凡德罗所走的道路。他们对空间的感觉是从立体主义那里继承的，但包豪斯却没有能够向着它本该有的方向去探索空间，但是赖特做到了。他过去和现在都不属于包豪斯学派，他比他们还要早得多。

I think. Shimonoseki? Yes. I stopped there, and the boat continued to Yokohama, but I did not stay on the boat. I took a train. I visited Kobe, Kyoto, and then Tokyo. Then, I went to Yokohama. So, I saw Japan before I saw the U. S. A.

And I stayed in Frank Lloyd Wright's hotel in Tokyo in 1935. Right. So, therefore, the excitement of seeing the world was such that I would run all kinds of risk. And there were risks, because the Japanese government could have stopped me at that time, you know. Do not forget that was 1935. (Mr.Shinich Okada in the audience laughs.)I guess you have not forgotten. So, therefore, it was a risky affair. Two of us decided we are going to do this. And we did not jump ship, but we asked for permission. And we said that if you allow us about 4 or 5 days, we'll meet you in Yokohama. we did it at our own risk, because the ship captain did not encourage us to do that. So, you see the excitement of seeing the world as such. And I could not resist. And after that, Honolulu, San

因此立体主义是我自己的选择，我想说的是就我自己的理解而言，它是比建筑本身更早的本源。你也可以认定另外一些较早的，或者是较晚的理念作为现代建筑的开端，但这是我的观点。我想说明的是，在对建筑产生的影响中，立体主义对格罗皮乌斯的影响不同于它对柯布西埃产生的影响，柯布西埃要更为接近立体主义的空间。

如果没有对"场所"的深刻理解，就不可能实现卢浮宫的金字塔

学生——贝先生针对格罗皮乌斯的国际式才是现代建筑的方向这一观点，提出历史和地域性的异议。对贝先生来讲，建筑表现与历史和地方性有什么样的具体联系？另外还有一个与此相关的问题，您有没有在什么方面受到日本的影响？

贝聿铭——关于历史嘛，唔，我并没有一种特定风格，对吗？我认为我没有。你们说我有，当然，我承认风格是有价值的。我有自己对风格的评价。我承认我们最终的确会形成特定的风格，但是我认为坚持一种风格是非常危险的，它会变成一种签名，你会变得过于松懈。

Francisco, then the new world. And it was a very precious moment in my life. And I still remember.

STUDENT:
You have mentioned that modern architecture starts from cubism. In what way did cubism influence Bauhaus and Le Corbusier?

I. M. PEI:
Cubism as understood by the Bauhaus people, is not the same as cubism as understood by Le Corbusier. I want to say that. Cubic space is much more evolved in Le Corbusier. That is, Gropius or even Mies van der Rohe. Ah, the sense of space is inherent in cubism, but the Bauhaus school has not explored space as much as should be. Frank Lloyd Wright, on the other hand, did. He does not belong, and did not belong. He was much older.

So, cubism is my own choice, I am saying it is the beginning, in my judgment, more than architecture. You can say it much earlier, you can say much later, but I choose that. But, I want to declare that within cubism as

对于一个建筑师来说，重要的是地点。正向我前面所说的，场所是地理环境的一种重要记号，它有自己的观点和方向。每个人都应该对此有所了解。每位建筑师都需要考虑这一点。他们应该了解这个场地，了解它的历史，它的根，而这种根是很难发掘的，但是一名建筑师如果想要有所创造，就必须对此有所发现。因为如果你对此无所作为，你的建筑就会枯萎。你不会感到兴奋，你也不会有太大成就。

因此，历史的重要性对我来说，就是那些特定地方的根。现在拿我在卢浮宫的项目作为例子，如果我没有对卢浮宫和法国的历史做过研究（实际我做了），我就不会获得成功，我可能早就失败了，我可能会被踢出法国，我几乎就要被踢出去了。（笑）但是我有能力保护自己，因为我研究过卢浮宫的历史，也研究过法国的历史，因此我就能够证明自己的设计，而不仅仅是说明自己的设计。它是和法国，巴黎，卢浮宫和谐一致的。很抱歉，太长了。（笑）

岸田——今天有几位从中国来的学生，今天的机会非常难得，你们如果有什么问题请提问。

学生——您到北京的时候我曾经听过您的讲话。您是什么时候开始认识到历史是重要的，或者说认识到根源是非常重要的？是从在哈佛大学学习的时候就开始

an influence on architecture, influence on Gropius is not the same as the influence on Le Corbusier. Le Corbusier is much closer to the cubic space.

Then, history. Ah, I do not have a style, or do I. I do not think I do. You say yes, but yet I value style. I think eventually a style will evolve. But there is a danger to stick with a style, it becomes a signature, and you loose a lot.

The important thing to an architect is place. As I said before, the site has a certain token of geographical significance, has a certain view and orientation. And that everybody would know. Every architect considers that. But, they know that site. There is history. Know the root. Those roots are very hard to find. But, an architect should find those roots in order to create. Because if you do not, architecture atrophies. You can not excite, you can not go too far.

So, therefore, importance of history to me is really the roots of that particular place. Now, my work at the Louvre, as an example, if I

认识到的吗？

贝聿铭——我在我自己的建筑实践中认识到了历史的重要性❶。在一开始的时候，我没有意识到历史的重要性，因为我所做的那些工程，如在纽约的低价住宅，它的历史就在那里，虽然并不是很长。接着，在像卢浮宫这样的项目中，我逐渐认识到，如果我不了解历史，我就不可能成功。

现在我自己挑选工作，因为我希望做一些有着有趣历史背景的工程项目，因为这会让我的建筑更为有意思。你们知道我想说什么。这是我后来才知道的事情。

强烈推荐旅行，旅行能将所学的东西变成自己的东西

贝聿铭——我将向大家推荐一样东西，这就是旅行。我在旅行中获得的知识要远比在学校中的所得要多。但是要从旅行中获益，就必须先学好历史，例如，如果你不了解文艺复兴的历史，或者是当你去雅典的时候，却不知道希腊的帕特农神庙，你就不能看到那些你本该看到的东西。但是如果你研究过文艺复兴的历史，当你去佛罗伦萨的时候，你的眼睛就能告诉你很多东西。

❶ 历史的重要性：虽然贝聿铭先生对格罗皮乌斯的现代主义风格提出了异议，但他的作品也并不是地域主义。不过，即使在"卢浮宫新美术馆"设计之前，贝聿铭的作品中也充分表现出了场所精神。他抵抗全球性的功能主义和合理主义，但并没有从传统风格和地域的东西中寻求解答。他在50年后向卢浮宫、巴黎和法国的历史学习，从中吸取营养精髓，与设计相结合，产生出新的设计。他从中国来到美国，作为建筑师经历了长年的岁月，才认识到并做到了这一点，这与一个老建筑师个人的经历是分不开的。

had not studied(which I did)the history of the Louvre, the history of France, I would not have succeeded then. I would have been defeated. I would have been kicked out of France. Almost was kicked out of France! (laughter) But I was able to defend myself, because I had studied the history of the Louvre, and the history of France. And therefore, I was able to justify my design, or rather define my design. That is, in tune with France, with Paris, with Louvre. Sorry, too long. (laughter)

I think I learn of this importance as I practiced architecture. In the begining, I was not aware of the importance of history, because the projects that I have to do like low cost housing in New York, a history is there, though there is not much. Then later on, as in assignments such as the Louvre, it was then I became aware, that if I do not know history I would not succeed.

Now I choose my work, because I want that particular project to have interesting history behind it. Because it makes my architecture

因此我们每天都能够看到历史的重要性,当每次我们在看那些建筑的时候,我想我们有时候没有必要去设计建筑。但是一定要去学习,你们必须了解历史。比起那些建筑杂志,我看得更多的是历史书籍。我也看建筑杂志,但是我确实更多地阅读历史书籍。我发现这是很值得的。而旅行是比教室中的学习更能开阔自身的方式。是的,要自己学习。

more interesting. You see what I mean. So, it is something that came much later.

I would recommend one thing to all of you, and that is traveling. I have learned more from traveling than I did in school. But to learn from traveling you have to study history first. Because if you do not know the history of the Renaissance, as an example, or history of the Parthenon in Greece, when you go to Athens, you do not see things that should be seen. But if you know the history of the Renaissance, then when you get to Florence your eyes are open to learn.

So, therefore, the importance of history is seen every day, every time you look at a building. It is not necessary to design of the building,I think. But just to learn, you have to know history. So, therefore I do more reading of history than of architecture magazines. I look at architecture magazines, but I do read history a lot. And I find it very rewarding, Then, traveling is more broadening than something you learn in the classroom. Yes. Learn yourself.

事务所刚成立时的佩罗

1998年11月20日（星期五）10:30am — 12:00pm　主持：大野秀敏

多米尼克·佩罗
DOMINIQUE PERRAULT

1953年佩罗生于法国中部的克勒蒙弗兰。1978年毕业于巴黎高等美术学院（Paris institute for advanced studies in the fine arts），成为法国政府承认的建筑师。1979年在国立土木大学学习城市规划，1980年在国立高等社会科学大学研究生院进行历史研究。1981年成立建筑师事务所。

1987年，佩罗在巴黎西郊的玛涅·拉·维列完成了他最初的作品：ESIEE工程技术学校。

1989年，他在"法国国家图书馆"（巴黎/1995）国际设计竞赛中获得一等奖，一跃成为国际建筑师。这个建筑也称为"密特朗图书馆"，是法国革命200周年之际巴黎"国庆工程"的最后作品，建于巴黎东部塞纳河左岸巨大的再开发地区，在长方形中庭的四周，有四栋L形高层玻璃幕墙建筑，给人非常深刻的印象。

1990年，他在巴黎的东南地区设计了一栋名为"伯利尔工业馆"的写字楼，四面也是用玻璃幕墙围起。现在，他的巴黎事务所就设在这栋楼里。

1992年，他在"柏林体育设施"设计竞赛中获得一等奖。在绿意盎然的长方形城市公园里，佩罗安排了圆形的自行车比赛场和长方形的游泳馆。这个设计无论从建筑的层面，还是从城市的层面都能看出自然环境与巨大的金属建筑物之间存在强烈的对比，同时又巧妙地调和共存，令我们感受到设计者的匠心。

（坊城俊成）

伯利尔工业馆/1990 © T.Bojo

柏林奥林匹克游泳馆和自行车比赛场
(Velodrom and Swimming Pool of Berlin)/1992
Photographer Werner Hutmacher

工程师家庭的环境
与绘画的结合点是建筑

大野——今天，法国建筑师多米尼克·佩罗为ＴＮ Probe❶的展览之事来到日本，我们有幸邀请他加入我们的这个系列，针对他所接受的教育给我们做一个讲座。

首先，我们想问一下，您从什么时候开始决定要把建筑作为自己的职业，它的起因又是什么？

佩罗——最初，我并不想搞什么建筑。（笑）

我出生在工程师家庭，与建筑没有什么关系。我并不知道建筑师是一种什么样的职业，也没有理由要学习建筑，成为建筑师。

我十四五岁的时候，正好是1968年法国的文化革命❷时期，从那个时候开始我知道了自己周围还有一个文化的世界，这个世界上还有技术以外的东西。

这样，我开始学习绘画，虽然画得不怎么样，可还是在十年中画了大量的画，当时自己认为是很好的作品。

可能是由于自己生长在工程师家庭的原因吧，周围总是有一种氛围，认为必须进行一些科学类专业的

❶ 创建于1995年日本的一个建筑与城市国际论坛。Ｔ指创建地Toriizaka（东京的鸟居坂），N指networking。与会者不拘于专业建筑师和城市规划师，也包括与之相关的其他专业的人士，以期不同领域之间的交流，从而更深入地思考城市和建筑的问题。
——译者注。

❷ 1968年法国的文化革命：由于这场文化革命，法国美术学院的建筑学院独立出来，成为一个新的建筑教育机构，行政归属上也从文化部转移到建设部，并改名为建筑学校。绘画、雕塑学科仍在美术学院。由于这样一个复杂的历史，建筑学校现在仍然常被叫做美术学院。

HIDETOSHI OHNO:
A Votre enfance déjà,vous vouliez devenir architecte?
DOMINIQUE PERRAULT:
En fait,ce qu'on peut dire,c'est que je ne voulais pas faire d'architecture.
Parce que je suis né dans une famille où il n'y avait pratiquement que des ingénieurs. Et dans cette famille,on ne connaissait pas l'architecture, on ne connaissait pas d'architecte:a priori il n'y avait aucune raison que je fasse des études d'architecte.
Lorsque j'ai eu 14-15 ans,c'était au moment des événements de mai 1968 en France,c'est-à-dire c'était la révolution culturelle française. A ce moment-là,j'ai découvert qu'il y avait finalement un monde culturel autour de moi et que le monde n'était pas qu'un monde technique.
Là,j'ai donc commencé à faire de la peinture. J'ai beaucoup peint,une peinture pas très bonne, mais j'en faisais beaucoup. J'ai peint pendant 10 ans. La peinture n'était pas bonne,mais elle était très sincère et très sympathique. Pour moi, pour tout le monde.
Le fait de peindre a créé évidemment dans ma famille un problème,puisque normalement je devais suivre la tradition,si on peut dire,et faire

学习。于是，我就考虑到，在技术与绘画的结合点上，有建筑这样一个专业，建筑一方面属于艺术的领域，另一方面属于科学的领域，当时我认为自己发现了一个很好的结合点。

于是，我就去了巴黎高等美术学院❸学习建筑。我是工程师家庭出身，故相比其他同学我有更多的数理知识的功底，因此，入学以后我很快就能同时考虑非常复杂的平面、立面、剖面等问题。而且，由于曾经立志学习绘画，当时就已经理解了绘图的技巧。

大野——那么我们是不是可以这样理解：您在选择专业的时候，并不是把建筑作为绘画和工程师家庭背景之间的一个妥协点，而是把对绘画的兴趣与家庭环境相结合，自然地考虑到建筑是实现自己理想的最好专业，才进行了这样的选择。如果当初没有遇到反对，佩罗先生是否会立志学习绘画呢？

佩罗——不清楚。不过我想我是不会选择画家的道路的。情况可能会更复杂。就像刚才讲的那样，进入美术学院以后，由于在一年级学生当中我是手头功夫很好的学生，所以从一年级开始就有老师邀请我到他的事务所去。这样，从一年级开始，在美术学院学习的同时，我就开始在老师的事务所里工作。在事务所里，我也受到一定的重视。所以说，这并不是是否把建筑

❸ 美术学院（ECOLE NATIONALE SUPERIEUR DES BEAUX-ART）：佩罗先生也不把它称为建筑分校，而是仍使用美术学院的叫法。在法国，要想成为政府承认的建筑师，就必须从这个建筑教育机构毕业。高中毕业后进入该校，经过5年的课程学习，再花一年的时间进行毕业论文和毕业设计，通过之后就可以毕业，并同时获得建筑师资格。

des études scientifiques.
Alors on a trouvé un compromis,un équilibre, en disant,voilà l'architecture,c'est pas mal,il y a un peu de technique,il y a un peu d'art. C'est une espèce de métier qui permet de rendre tout le monde content,chacun y trouve un peu ce qu'il veut,un peu n'importe quoi. Et donc c'était une bonne base d'accord entre les parents et les enfants.
Je donc suis allé à l'école des Beaux-Arts. Avec un bagage,si je puis dire,technique qui était très solide puisque j'avais fait des études mathématiques et techniques. J'étais capable de faire la relation entre la coupe,le plan,la façade avec des objets mécaniques;donc des coupes assez complexes. En même temps que j'avais ce bagage technique très solide,je savais dessiner,et surtout je voulais peindre. Il y avait donc là une espèce de mélange qui était en fait pas si mal pour, petit à petit,devenir architecte.
HIDETOSHI OHNO:
Si c'était possible,vous auriez pris le chemin vers la peinture?
DOMINIQUE PERRAULT:
Je ne sais pas. Je ne crois pas. C'est plus compliqué que ça:lorsque je suis entré à l'Ecole des Beaux-Arts,tout de suite,je savais dessiner, je dessinais normalement,je construisais

当作一个妥协点的问题，而是当时我就已经成为建筑师了。

从一年级开始就一边在美术学院学习，一边在事务所参加实际工作

大野——从一年级开始就在事务所里工作这种情况，在法国的建筑学生当中是常有的事吗？

佩罗——我想从一年级开始就两方面同时开始的学生不多。学习一段时间之后再开始打工的情况是有的。

大野——佩罗先生受到那个老师的邀请，对您来说是一个机会。那么您自己是否也对参加实际工作感兴趣呢？

佩罗—— 是的。我认为，建筑学习并不能只限于学校的学习。学校的学习，主要是了解建筑和一些初步的学习，而对建筑的理解和学习应该是在实际工作中进行的。

大野——在那个事务所中，您还是一个一年级的学生，又没有什么知识，您做了哪些工作呢？

佩罗——当然首先是晒蓝图（笑），最初就是做的这些打下手的工作。学会描图之后，就开始做模型。不过我并不认为做模型是打下手。做模型，就是制作，是实际建设过程中非常重要的一环。模型是根据图纸制

normalement.
Ce qui en général pour un élève architecte en première année est assez rare. Je n'ai donc jamais eu de problème avec la construction.
Alors,tout de suite,mes professeurs ont vu que j'arrivais à dessiner deux trois choses,et tout de suite j'ai travaillé en dehors de l'école chez mes professeurs architectes,et j'ai toujours travaillé chez mes professeurs. Ils trouvaient finalement que j'étais un élève pas mal,qui était pas cher, qui dessinait pas trop mal,et c'était une très bonne chose.
Tout de suite,j'ai travaillé en plus de l'école et là j'ai appris mon métier,car finalement,ce métier,ce que je dis souvent,c'est le temps de l'école plus dix ans de pratique. Après,on peut commencer peut-être à dire qu'on est architecte.
HIDETOSHI OHNO:
Pour un élève d'architecte en première année, est-ce qu'il est normal de travailler à l'agence?
DOMINIQUE PERRAULT:
Non c'est pas très normal.
HIDETOSHI OHNO:
Vos professeurs vous ont fait travailler à l'agence, n'est-ce pas?
DOMINIQUE PERRAULT:
Le professeur m'a dit,toitu viens travailler chez moi. Voilà. Mais je crois que l'architecture ne

作的，我能够一边做模型一边学习平面图、立面图，并学习结构图纸的识图。对我来说，做模型是一个非常宝贵的经历。

大野——大学里不教这些吗？

佩罗——在学校里做的模型和在事务所里做的模型不一样。虽然在学校里做的模型对于建筑理论等课程的学习也非常重要，但是，事务所的模型是实际上要建成的东西，或者将来可能要建成的东西，这对我来说更重要。进一步来讲，两者同时做也是一个非常好的经历。我35岁时在"法国国家图书馆"设计竞赛❶中获胜，我想这应该归功于在这之前十几年的积累。仅仅是学校或仅仅是事务所的学习都是不完善的。

大野——今天在座的同学中一定有许多人将来也希望成为建筑师，那么佩罗先生是否也劝大家采用这个方法呢？

佩罗——我想，这并不是一个简单的劝和不劝的问题。打一个比方，请大家想一下，往楼上去的话，是使用电梯还是使用楼梯？不能说使用了电梯就能够一下子成为建筑师，事情并不是这样。应该像上楼梯一样一步一步地向上攀登，连续不断地努力，才有可能达到目标，我想应该是这样吧。

❶ "法国国家图书馆"设计竞赛：1989年举行的国际设计竞赛，有220个单位参加，选出了包括J.斯特林在内的5个优秀方案。最后，当时的密特朗总统任命佩罗进行设计。另外，特别奖中有努维尔和库哈斯的方案，后者引起许多议论。

s'apprend pas dans une école. Ce n'est pas en allant à l'école qu'on apprend l'architecture. Lorsqu'on va à l'école, ça permet d'avoir un sentiment, une idée, mais apprendre l'architecture, c'est un parcours initiatique. L'expérience de l'architecture, c'est construire. Tant qu'on n'a pas construit, on ne sait pas ce qu'est l'architecture.
HIDETOSHI OHNO:
Qu'est-ce que vous avez fait à l'agence?
DOMINIQUE PERRAULT:
J'ai fait des tirages de plans, des photocopies. C'est très très formateur. C'est l'école de base. Et petit à petit j'ai surtout fait des maquettes. Faire des maquettes, ça a été très important.

Parce qu'une maquette, ça se construit. C'est certainement l'élément qui est le plus juste lorsqu'on travaille un projet, et savoir faire une maquette, c'est savoir lire un plan, savoir lire une coupe, savoir comprendre une structure... La maquette est une espèce d'épreuve de vérité.
HIDETOSHI OHNO:
A l'école, on ne vous apprend pas cette choselà?
DOMINIQUE PERRAULT:
Si. Les maquettes que l'on fait à l'école sont des maquettes qui sont très intéressantes, mais qui sont pas des vraies maquettes. Parce que lorsqu'on travaille dans un bureau, et qu'il y a un vrai projet, la maquette, elle doit être juste. Il

大野——看来您的学生时代是在繁忙中度过的。除此之外,作为一名以建筑师为目标的学生,您还做了些什么事呢?例如旅行什么的。

佩罗——工作。

大野——从早到晚都在工作?

佩罗——从早到晚都在工作。(笑)

大野——法国的建筑学校需要学习多长时间?

佩罗——六年。

大野——那么,六年您都是这样度过的吗?

佩罗——是的。

研究 19 世纪后半期巴黎区政厅的毕业研究受到好评

佩罗——当然学习的过程并不仅仅是六年,而是一直持续至今。我毕业设计的时候跟随了两位老师,一位是安德鲁·葛兰巴克❶老师,他是一位理论家;一位用分析的方法研究城市的研究者。在他的指导下,我对19世纪的城市进行了分析,作了一些历史性的、理论性的研究。另一位老师是德国人,是一位在实际工程中进行大型工程设计的建筑师,他用粘土等材料来做模型。我分别在从事理论研究和实际设计的两位老师

❶ 安德鲁·葛兰巴克(Antoine Grumbach,1942—):1967年美术学院毕业后,进入国立高等研究院。1968年遇到美术学院解体,1970年开始成为UPA6(现在的拉·维赖特建筑学校)任教授。佩罗学生时曾在他的事务所里工作。

y a une réalité.
C'était comme un très bon équilibre entre un travail de recherche que l'on peut faire à l'école et un travail très opérationnel,très pragmatique, très concret que je faisais dans un vrai bureau d'architecture.
Et c'est cela qui a été formidable. C'est pour ça, on ne peut pas... bon il y a le destin,... mais on ne peut pas construire de très grands bâtiments, comme j'ai pu faire,à 35 ans,sans avoir derrière 20 ans de réflexion,de travail,et théorique et concret. C'est pas possible. C'est pas un miracle.
HIDETOSHI OHNO:
Est-ce que vous conseillez aux élèves d'architecte de faire la même chose que vous?
DOMINIQUE PERRAULT:
J'avais un professeur qui disait:ne prends pas l'ascenseur,monte par l'escalier. Cela voulait dire monte marche après marche.
Il faut prendre l'escalier,parce que c'est un parcours. Et marche après marche,on apprend. Alors que si l'on prend l'ascenseur,on est à l'école,on a un diplôme et ensuite on croit qu'on est architecte,mais en fait,à ce moment,tout commence.
HIDETOSHI OHNO:
En travaillant à la fois à l'école et à l'agence, vous avez fait d'autres choses pour devenir

的指导下研究了19世纪的城市,这对我来说是非常有益的。

大野——提到毕业设计的话题,那么我想问一下,您的毕业设计都做了哪些工作?

佩罗——研究了19世纪后半期巴黎的区政厅。巴黎有20个区政厅。19世纪后半期是巴黎城市扩张的时期,在大约50年的时间里,新建了11个区政厅,改建了9个,变化非常大。我就是对这20个区政厅进行了建筑方面的分析。在我做这个研究之前,还没有人对这20个区政厅建筑进行全面的比较研究,甚至没有它们的图纸。也就是说,对于这些分散在各处的20个区政厅,没有一个从建筑的角度或者从城市的角度上进行的研究。

大野——这些研究最后归纳成一个设计方案了吗?

佩罗——没有。我并不是选择一个区政厅进行实际设计,而是针对20个区政厅和巴黎的关系,从城市的层面上进行分析。我把这作为研究课题。至于实际的设计嘛,我在事务所从早到晚都在做模型,已经不愿意再做设计。(笑)

大野——在同一个年级中,用一个研究性的课题作毕业设计,是不是比较特殊?

佩罗——大部分同学都提交了一个普通的方案。我是

---

architecte;par exemple,voyager etc. ?
DOMINIQUE PERRAULT:
J'ai travaillé.
HIDETOSHI OHNO:
Du matin au soir?
DOMINIQUE PERRAULT:
Et du soir au matin aussi.
HIDETOSHI OHNO:
En France,normalement pendant combien d'années on étude à l'Ecole d'Architecture?
DOMINIQUE PERRAULT:
6 ans.
HIDETOSHI OHNO:
Durant ces six ans,vous travailliez toujours à l'agence?
DOMINIQUE PERRAULT:
Oui. Et ensuite aussi. Ca ne s'arrête pas. Cela ne s'arrête jamais.
Par exemple,l'année de mon diplôme,j'avais deux professeurs importants. J'avais d'une part un professeur qui s'appelle Antoine Grumback,qui est un professeur plutôt théoricien,qui fait un travail d'analyse urbaine sur la ville. Et j'avais un autre professeur,un professeur allemand qui lui était un grand professionnel qui construisait des grands bâtiments:avec ce professeur allemand,je faisais des maquettes en pâte à modeler,et avec André Grumback,je faisais des

进行了理论分析。结果我的论文发表在关于建筑和城市的杂志上，在大学内外得到好评，大家都认为是一个非常好的研究。

为了研究20世纪巴黎周围的新城市，进入国立土木大学

大野——在这之后，佩罗先生进了国立土木大学❶学习。

佩罗——为什么当时我认为区政厅的研究是重要的呢，这是因为在建造一个建筑的时候，当然同时建造几个建筑的时候也是一样，怎样处理建筑与周围环境的关系这个问题非常重要，我对这一点非常感兴趣。我认为能够考虑这些事情，是成长为建筑师的重要一步。建筑并不是把一个单体建筑建成之后放在那里就行了，必须要考虑和周围环境的关系。以刚才区政厅的研究来讲，如果要在分析20个区政厅的基础上，提出一个新的方案的话，那么，巴黎总共散布了20个区政厅，就必须以区政厅的设计为目的对巴黎进行整体的考虑，首先提出一个城市规划的方案来。因此，从考虑周围环境这个意义上来讲，我认为对区政厅的研究是一件非常有意义的事情。

❶ 国立土木大学（Ecole Nationale Superieur Des Ponts Etchaussee），并不是建筑，而是国立的土木和城市规划的高等教育机构。该校与美术学院不同，如果直译，其名称应为国立桥梁和道路的教育机构。

analyses du tissu urbain(urban fabric)des villes au 19ème siècle.
En même temps,il y avait ces deux pratiques, qui sont pour moi tout à fait complémentaires: théoriques,historiques et complètement opérationnelles.
HIDETOSHI OHNO:
Qu'est-ce que vous avez fait en tant que votre diplôme?
DOMINIQUE PERRAULT:
J'ai fait un diplôme sur les mairies annexes de la ville de Paris:j'ai fait une étude sur la construction des mairies dans la deuxième partie du 19ème siècle.

Dans la deuxième partie du 19ème siècle,Paris s'agrandit. Paris qui a 20 arrondissements aujourd'hui a annexé les villes autour au milieu du 19ème siècle. Alors 11 mairies ont été construites,et les autres mairies ont été restructurées et rénovées. En 50 ans,les 20 mairies de Paris ont été transformées ou ont été construites. J'ai fait un travail comparatif d'analyse entre ces 20 bâtiments sur cette période de 50 ans à peu près. Il n'y avait aucun livre écrit sur ces bâtiments,il n'y avait aucun plan qui compare l'ensemble des 20 mairies,il n'y avait aucune réflexion urbaine,architecturale sur ces bâtiments,qui en fait,ont organisé les quartiers

从美术学院毕业后，我进入了国立土木大学这所土木工程大学学习。我的学习，在建筑的层面上，尤其是在城市的层面上，与美术学院的毕业设计中进行的巴黎二十所区政厅的研究相连贯。在法国，随着共和国的成立，区政厅也同时出现。这并不是一个偶然的产物，对巴黎城市来说这是一个本质性的变化。

现在的状况是，我们不得不在巴黎的周围建造新的城市，这就碰到这样一个问题：用前一个世纪的方法来考虑新的城市的建设是否合适？在国立土木大学，我研究了新城市的建设及其形成的过程。我对19世纪和现在的区别很感兴趣。

在美术学校，我研究学习了19世纪的巴黎，在国立土木大学，我研究学习了现在的巴黎，或者说是对巴黎周围的城市进行了学习研究。

大野——是不是因为在美术学院没有学到现代城市中关于新城市问题方面的知识，以及城市规划等方面的知识，所以想在新的学校中进一步学习呢？

佩罗——美术学院也有城市规划的课程，我想知道的，并不完全是城市规划课程中的内容，而是一个城市究竟是怎样形成的？不是怎样从建筑逐渐形成城市，而是一个城市的框架是通过什么样的战略性策略来形成的，我对这一点感兴趣。比起建筑是怎样建造的这一

autour d'eux. Et j'ai voulu travailler sur cette relation entre l'institution publique et la ville.
HIDETOSHI OHNO:
Avez-vous dessiné concrètement ce projet de diplôme?
DOMINIQUE PERRAULT:
Non. Concrètement... Concrètement, c'est pas concret. C'est un projet théorique, c'est un projet qui... Je ne voulais pas faire de projet concret, je faisais des projets concrets tous les jours. Donc, j'allais pas encore faire de la pâte à modeler.
HIDETOSHI OHNO:
Est-ce qu'il était normal de faire un projet théorique par comparaison avec des diplômes d'autres étudiants?
DOMINIQUE PERRAULT:
Je sais pas si c'était normal, mais c'est un diplôme qui a été publié dans les journaux et dnas les magasines... C'était un bon diplôme.
HIDETOSHI OHNO:
Les autres étudiants ont-ils fait des projets concrets?
DOMINIQUE PERRAULT:
Oui plutôt. En fait, ce diplôme sur les mairies a été très très important pour moi. Et aussi en peu dans le monde de l'architecture en France, parce que c'était une des premières fois qu'on mettait en relation l'architecture

问题而言，我对为什么建造建筑或者城市这一类问题更感兴趣。

大野——原来是这样。你想知道的，并不是19世纪那样由建筑而形成的城市，而是现代的城市战略问题，或者是城市经营问题。

佩罗——是的。

大野——在那里学习了一年，在这段时间里您过的是什么样的生活呢？

佩罗——这段时间仍然继续在事务所里工作。

大野——一直是同一个事务所吗？

佩罗——不是。有设计竞赛之类的事情的时候，我会连续三个月在一个事务所工作，有别的项目的时候，我又会到别的事务所去，并不是一直在一个事务所里工作❶。也并不是我一个人这样，其他人也一样。

❶ 并不是一直在一个事务所里工作。佩罗先生用了"豪华生活的奴隶"这样一种特殊的表达方式，他3个月呆在事务所埋头工作，等稍微有点钱了就回到学校用于学习，然后又几个月在事务所工作，重复着这样一种生活方式。

❷ 国立高等社会科学大学（ECOLE DES HAUTES ETUDES EN SCIENCES SOCIALES）：正如佩罗先生所言，这并不是一个培养建筑师的学校。一般来讲，这是一个攻读社会学科专业博士学位的教育机构，也就是只有研究生院的大学。

为了研究18世纪修道院和巴黎的关系，进入国立高等社会科学大学研究生院学习

大野——从国立土木大学毕业之后，您又进入国立高等社会科学大学❷研究生院学习历史，这又是出于什么样的原因？另外，您又研究了哪些历史？

et la fabrication d'un quartier.
Ce qui était intéressant dans ce diplôme, c'est que je n'ai pas étudié l'architecture, et c'est ce qui me paraît le plus important : j'ai étudié les relations entre les architectures, et ça, je pense que c'est vraiment architecte.
HIDETOSHI OHNO :
Après l'Ecole des Beaux-Arts, vous êtes allé à l'école des Ponts et Chaussées, n'est-ce pas?
DOMINIQUE PERRAULT :
Ensuite lorsque j'ai fait le travail à l'Ecole des Ponts et Chaussées, je voulais essayer de comprendre les relations entre les choses que j'avais étudiées durant cette période très importante, puisque c'est la mise en place de la République : c'est donc une période historiquement très importante, l'histoire des mairies, c'est la République.
C'est pas un bâtiment par hasard, c'est un bâtiment essentiel. Je voulais savoir lorsque l'on construit une ville nouvelle autour de Paris, s'il y a les mêmes idées qu'au 19ème siècle?
Donc, aux Ponts et Chaussées, j'ai étudié la construction, la réalisation, le process de réalisation d'une ville nouvelle en France, cela m'a intéressé de comparer une période contemporaine et une période du 19ème siècle.
Ce fut le prolongement de l'étude sur ces mairies.

佩罗——我从美术学院，到国立土木大学，再到国立高等社会科学大学研究生院，这期间一直有一个连贯的东西，也就是刚才我提到的"为什么"、"怎么样"这一问题。对我来说，怎样建造建筑、怎样成为建筑师这个问题不重要，我对为什么要建造建筑、为什么要成为建筑师这个问题感兴趣，这也是我进入国立高等社会科学大学研究生院的原因。

我认为，怎样建造建筑，例如如何选择材料，自己应该拥有什么样的风格，等等，这都是非常个人的问题。我对这些问题不感兴趣。我感兴趣的是，为什么要建造，或者为什么自己要成为建筑师，比起个人的问题来说，这更带有团体性质——为什么20世纪要建造这样的建筑，我认为这与团体性，或者说民族意识等问题有关。

在国立高等社会科学大学研究生院，我研究了18世纪的修道院。在18世纪，修道院与城市街区之间有着非常密切的关系。在20世纪，也应该有与20世纪新的街区相符合的东西。我对这件事情感兴趣，所以研究了18世纪的修道院。这个学校是一个非常有名而且很大的学校，但它是培养博士的学校，并不是培养建筑师的学校。

大野——这么说，你在那里度过的是以学问为中心的

HIDETOSHI OHNO:
A l'Ecole des Beaux-arts, on n'apprend pas des villes nouvelles?
DOMINIQUE PERRAULT:
Si, bien sûr, on apprenait des villes nouvelles, on parlait des projets. Mais moi, ce qui m'intéressait, c'était la stratégie, l'organisation, comment le pouvoir organise la ville et fait construire tel et tel bâtiment...Ce qui m'intéressait, c'était pourquoi l'architecture est construite, et pas comment?
HIDETOSHI OHNO:
C'est à cause de cela, êtes-vous allé à l'Ecole des Ponts et Chaussées?

DOMINIQUE PERRAULT:
Oui.
HIDETOSHI OHNO:
Est-ce que vous avez travaillé toujours à l'agence?
DOMINIQUE PERRAULT:
Oui.
HIDETOSHI OHNO:
Avez-vous toujours travaillé à la même agence?
DOMINIQUE PERRAULT:
Non, c'était par périodes, par exemple pour faire des concours. On n'applle ça dans le langage des architectes "un nègre de luxe". Je m'explique: un nègre, c'est quelqu'un un peu comme un

生活了。

佩罗——还和以前一样在事务所工作。（笑）在法国，一般来说，越是进行高学位的学习，必须要听的课就越少。主要是自己进行研究，所以有更多的时间去事务所。

大野——这个关于修道院的研究是否以书的形式归纳出版了呢？

佩罗——没有出书。

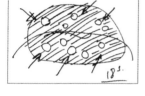

1

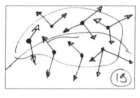

2

如果讲到关于修道院的研究的话，是这样的，（一边画图）这是巴黎，修道院就像这样分布在这里。在18世纪的巴黎，就是这样高密度地集中在一起。图中的圆点代表教会和修道院制造的一种虚空间。可以看出18世纪的社会体系基本上是以宗教权威为中心的。巴黎的街道也是这样的状态。到了18世纪末，社会进入了法国革命时期，像箭头所示的这样，开始从各个方向像巴黎的中心涌动（图1）。

进入19世纪，共和国成立，这些点代表区政厅（这不是虚空间）。在共和国的旗帜下，如箭头所示，迅速地向周围扩张（图2）。这并不像宗教那样向中心汇聚，而是向周围扩散。现在的巴黎就是这样，现在我们在巴黎见到的宽敞的道路基本上都是这个时代修建的。

esclave. C'est une expression pour dire que l'on arrive comme un mercenaire.
Je viens pendant trois mois et je vais travailler sur un projet. Un certain nombre d'architectes fonctionnent comme cela. Ca me permettait de travailler deux mois et ensuite d'avoir un peu d'argent,et de continuer mes études.
HIDETOSHI OHNO:
Après l'Ecole des Ponts et Chausées,en plus vous avez fait l'histoire à l'Ecole des Hautes des Sciences Sociales,pourquoi l'histoire?
DOMINIQUE PERRAULT:
Tout ceci est dans le même prolongement. C'est-à-dire,c'est la question que je posais tout à l'heure:pourquoi et comment?
Les architectes,et surtout quand on est étudiant, se posent la question:Comment je vais construire ? Comment je vais être architecte?Je crois que c'est exactement la mauvaise question. La question est: Pourquoi je vais construire et pourquoi je vais être architecte. C'est pas du tout la même chose. Il y a deux dimensions. La question du comment est une question personnelle,une question de style,mais c'est surtout une question de sensibilité. Comment je suis architecte?Certains vont construire avec certains matériaux dans un certain esprit,qui leur est personnel,intime et qui est lié vraiment à la personne.

在理论与实践之间
来回走动是最关键的

大野——可以说佩罗先生从学生时代开始整个二十多岁的时光是没有虚度的,您能够对自己的将来进行准确地判断,明白今天应该做什么,明天又应该做什么,每一步都有非常明确的目标,并赋予行动。我想问一下,这么明确的判断全都是你自己思考决定的吗?中间没有一点犹豫吗?

佩罗——刚才我讲到一个历史的过程,但这并不是历史学家所考虑的那种"过程",我是用地理的、平面的角度来进行研究的。历史学家会说18世纪是这样,那么19世纪就会是那样,20世纪又会是另外一样,强调时间流程中每一段的分割。我并不是这样,我从地理的角度进行学习,所以我的视野、我看问题的方法,有一个连贯性。

大野——进行了城市规划方面的研究之后,一般来说,具有实际设计经验的学生都非常希望能够尽快地开始实际工作,把自己设计的建筑建造起来,但佩罗先生却能在年轻的时候就想到将来,进一步去学习历史,这一点给我非常深的印象。你是否压抑了自己立即建

Alors que la question du pourquoi est une question d'éthique,est une question collective. Et c'est la question certainement la plus importante pour les architectes.
Ce qui m'intéressait dans ces études,c'était de travailler toujours sur des bâtiments publics ou institutionnels d'une époque.
Et dans cette école,qui est une très grande école française,très célèbre,puisque c'est l'école qui permet de devenir docteur… J'ai étudié les couvents. Les couvents. C'est l'institution qui représente au 18ème siècle à Paris,l'église. C'est pas l'église. (shudoin)
Et donc j'ai étudié les couvents au 18ème siècle.
Voilà. Ce qui était intéressant,c'est que les couvents organisaient la ville au 18ème siècle, les mairies organisaient la ville au 19e siècle. Et les villes nouvelles organisaient la ville au 20ème siècle.
HIDETOSHI OHNO:
A cette époque également,avez-vous travaillé à l'agence d'architecte?
DOMINIQUE PERRAULT:
J'ai toujours ponctuellement travaillé. En France,c'est plus facile de travailler lorsqu'on avance dans les études,parce qu'on a moins de cours. On a des rendez-vous avec un professeur: C'est un travail beaucoup plus de recherche et

造建筑的欲望，或是有别的什么想法……？

佩罗——您说得很对。但是比起建造的欲望来讲，我更多的时候是在思考理论与实践的平衡。

大野——您认为对于一个建筑师来讲这很必要吗？

佩罗——我认为在理论与实践之间应该不停地来回走动，这一点非常重要。

大野——是这样，我明白了。

在20世纪的"法国国家图书馆"中制造虚空间

大野——一个月前我走访了"法国国家图书馆"，我感到建筑中庭中的虚空间❶具有强烈的纪念性，这一点让我非常感动，今天听到您的讲话，知道您在学生时代进行了许多研究，其中的成果都结晶在那个设计中，让我深受启发。

佩罗——在"法国国家图书馆"的设计中我融合了法国18世纪和19世纪的特点和形象。18世纪的特点体现在图书馆的读书空间中，是一个修道院的空间，一个寂静的要素，一个与城市隔绝的空间。19世纪的空间特点也表现在国家图书馆中，我的想法是，图书馆是一个信息发源地，而且这是国家图书馆，更应该是

❶ 虚空间：从词的本意上来讲是"什么都没有，空虚，空间"的意思，从建筑和城市的意义上来讲，并不是指什么都没有的剩余空间，而是刻意制造的没有建筑结构存在的空间。

on peut travailler beaucoup plus facilement. C'est comme un travail de post-graduate.

HIDETOSHI OHNO:
A cet établissement, avez-vous fait un projet? Ou ce que vous y avez fait a été publié?

Non. Parce qu'il n'y avait aucun dessin. C'étaient uniquement des études sur archives, qui étaient des textes. C'était un travail de recherche universitaire.

J'ai construit la bibliothèque nationale, qui est le croisement de cette analyse du 18ème siècle et du 19ème siècle. Je peux expliquer cela. C'est très intéressant.

Vous n'avez pas un morceau de papier, pour qu'on puisse dessiner quelque chose?
Je vais expliquer. Par exemple, voilà Paris: l'histoire des couvents. Ça c'est le 18ème siècle. Au 18e siècle, la ville est très dense, il y a des petites rues, des petites maisons etc. Et les vides, ce sont les églises et les couvents. Au 18ème siècle, la densité c'est comme ça. (fig.1) C'est la religion qui est un ordre relativement secret, ou tout au moins particulier. Il organise la ville autour d'un système qui est en fait le système des églises et des couvents, et qui est un certain type de pouvoir, le pouvoir d'ordre religieux.

Au 19ème siècle, c'est la révolution à la fin du 18ème siècle. Ça c'est le 19ème siècle, et c'est

一个公共建筑，比起封闭的空间来说，它更应该是一个具有交流性质的交流场所，这也是19世纪空间的特点。现在，大家应该能理解为什么我会在一个建筑中设计18世纪和19世纪两种特性的空间了。

另外，我想讲一下，我不是建造"建筑"，而是建造"场所"。虽然现在大家会把一些东西看成是"建筑"，但是，时间会制造"场所"，因此，我想，我建的图书馆，也会随着时间的流动而出现新的东西。

这个图书馆是把"虚空"作为一个中心要素进行设计的。所谓"虚空"，并不是"没有"的意思，"虚空"在城市中具有"存在"的意义，我想这是很有趣的事情。

我有意设计了大野先生所说的"虚空间"的部分。那么，设计这个"虚空间"的意义何在呢？我们知道，那个地区❶现在什么都没有，也没有大型的工程，可以说是在一个虚空间中设计了虚空间。但是，等几年后周围的建筑建起来的时候，最后留下来的就是我所建造的"虚空间"了。因此，在现在这个时间里，说那个图书馆的虚空间没有意义是不正确的，我建造的是"明天的虚空间"。也正是为了向大家说明这一点，今天我才来到了这里。（笑）

法国国家图书馆
Photographer Georges Fessy

❶ 那个地区：即巴黎东部赛纳河左岸的巨大开发地区，对岸有"贝尔西公园"。

complètement l'opposé,puisque c'est la république:la démocratie commence. C'est à partir des mairies,de ces points,que l'ordre de la ville va être organisé. Et le Paris que vous connaissez,c'est celui-ci. Les grandes avenues,les institutions,les parcs,tous les grands axes,tout le système urbain de Paris est un système qui est lié à la mise en place de la république. (fig. 2) L'idée de la bibliothèque nationale,c'est un peu le croisement du 18ème et du 19ème siècle. C'est en même temps un couvent,un cloître, c'est-à-dire un lieu calme de méditation pour la lecture,donc un lieu fermé,un lieu un peu en dehors de la ville.

Et cela doit être en même temps un lieu ouvert sur la ville,puisque c'est un bâtiment public,et que la culture doit être accessible à tout le monde:il faut que ce bâtiment soit et fermé pour pouvoir travailler dans la sérénité,et ouvert sur la ville pour pouvoir échanger avec la ville l'activité et l'ouverture.

Il faut dire que ce bâtiment,n'est pas un bâtiment,c'est un lieu. Et il faut avoir la patience et l'humilité d'attendre que le temps passe pour que ce lieu devienne un lieu dans la ville. Aujourd'hui la bibliothèque,c'est un vide,c'est clair.Mais c'est un vide au milieu d'un vide. Parce que le quartier n'est pas encore construit.

为什么建造建筑、怎样建造建筑，这是一个事物表和里的两个方面

安藤——听到刚才的讲话，我感到非常重要的一点是，我们在思考建筑的时候，很容易陷入怎样建造建筑这样的思维定势中去，而重要的应该是为什么建造。正因为不思考这个问题，我们日本的建筑才总是受到批判。我想，从现在开始，我们也必须认真思考为什么建造这个问题。

佩罗——确实是这样。进一步讲，怎样建造和为什么建造这两个问题，也不能简单地讲谁更重要，虽然说为什么建造是很重要的问题，但这就像铁轨一样，怎样建造和为什么建造是一组铁轨中的两条，无法分开，必须在这两个问题之间不断地来回走动才行。

与18、19世纪不同，20世纪的问题变得非常复杂。对一件事情，我们不能简单地判定是好还是坏。我想现在是一个要同时解答复杂而且是复数的问题的时代。

（一边画着图）这是建筑工地（图3）。有两个人站在这儿，两个人都非常幸福，这时建筑登场了，立即就在两人之间建造了一堵墙，这堵墙把两个人分开了，这样，就会有一个为什么建造这堵墙的问题。那么，为

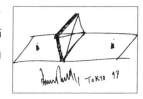

3

Mais dans 10 ans, dans 20 ans, dans 50 ans, autour de ce vide de la bibliothèque, il y aura des milliers de gens qui vont vivre.
Donc, le seul vide qui restera, c'est le rectangle de la bibliothèque. Et c'est pour ça qu'aujourd'hui on dit:ah!ça ne marche pas. C'est vide etc. . C'est vrai aujourd'hui. Mais aujourd'hui c'est rien dans l'histoire de la ville. Demain…
HIDETOSHI OHNO:
En votre curriculum vitæ, on dirait qu'il n'y a aucun détour, est-ce que vous avez eu quelqu'un qui vous a indiqué le chemin?
DOMINIQUE PERRAULT:
Non, ce que j'ai décrit, ce sont des situaitons historiques, c'est une analyse de l'histoire sous un certain angle ou avec un regard de géographie. C'est-à-dire que ce n'est pas une étude de l'histoire comme le fait un historien, c'est plutôt l'étude de l'histoire, des couvents, des mairies, des villes nouvelles, mais avec un regard de géographe.
C'est une vision tout à fait différente de l'architecture:Un historien qui étudie l'architecture, il étudie le style, il étudie des relations qui sont finalement assez ponctuelles et assez factuelles;Un géographe, lui, va étudier un territoire, et donc, il va étudier les relations entre des événements, entre des

什么建造呢，就是为了要分成两个空间。这就是为什么要建造的说明。

接下来，我们再谈一下怎样分成两个空间的问题。例如，用石头分成两个空间。怎样建造的问题就是要考虑用什么材料等问题，用石头或者玻璃或者木材。再接下来就是透明还是不透明的问题。怎样建造建筑就是这样的问题，和为什么建造建筑是一个一体的问题。

安藤——刚才提到"法国国家图书馆"，对这栋建筑我也很有印象。实际上在竞赛的时候，前几天来过的贝聿铭先生曾经邀请我做评委，但由于时间不巧，我只是在评审之前看了一下全体参赛作品。在设计竞赛这种场合，就像刚才所说的那样，理论和表现完美结合的方案，是非常有说服力的。在当时看到的几个作品中，我认为佩罗先生的方案是最有说服力的。

听了刚才的讲话，我再次体会到大野先生所讲的中心部分的虚空间的作用，还有四周楼梯状的建筑部分对周围环境产生的影响等方面，从建成后的效果来看，是非常合适的。

今天的谈话有一点让我感触很深，就是佩罗先生同时兼有理论研究和实际表现两方面的能力，并不偏重任何一方——日本的学生往往会偏向其中一方——

situations, entre des lieux. Et certains de ces lieux, sont des bâtiments. Et c'est le regard qui donne cette cohérence, c'est un regard de géographe, c'est pas un regard d'historien.
HIDETOSHI OHNO:
Vous avez fait des études dans ces trois établissement et en même temps vous avez travaillé chez architectes, pendant ce temps-là, est-ce que vous avez pensé que vous aviez envie de réaliser votre projet?
DOMINIQUE PERRAULT:
C'est l'équilibre entre la théorie et la pratique.
HIDETOSHI OHNO:
Pour devenir architecte, il faut avoir cet équilibre?
DOMINIQUE PERRAULT:
L'équilibre, non. L'équilibre, ça veut dire égal, donc c'est pas nécessaire ça. Ce qui est nécessaire, c'est l'interactivité, la relation, la synergie entre la théorie et la pratique. De passer de l'un à l'autre, de pouvoir traduire concrètement une idée théorique, de pouvoir aussi analyser théoriquement une idée concrète. C'est-à-dire le fait de pouvoir passer de l'un à l'autre.
HIDETOSHI OHNO:
Le pourquoi est le plus important? Ou alors, Pourquoi est plus important que Comment?

建筑师必须不断地在两者之间来回走动,并培养自己丰富的表现力。

正好现在我和佩罗先生都在参加"法国国家图书馆"前面的桥的设计竞赛,两个竞赛对手在这里聚在一起,让人有一种不可思议的感觉。(笑)佩罗先生在"法国国家图书馆"一举成名后,又在许多设计竞赛中获胜。我觉得这期间设计水平提高很快。对年轻人来说,如果连续不断地进行挑战,在十年当中,说不定就会拿到非常大的项目。听了佩罗先生的谈话,我感到不停地挑战是一件非常重要的事情。

音像图书馆
(Media Library of Véissieux)
Photographer Georges Fessy

DOMINIQUE PERRAULT:
C'est pas une question de valeur. Pourquoi n'est pas plus important que Comment. Ce qui est important,c'est de se poser la question du pourquoi,car il n'y a pas de hiérarchie.
Moi,je pense que c'est un travail comme un faisceau,comme un réseau. C'est-à-dire que si l'on se pose que la question du comment,c'est pas suffisant. Si l'on se pose seulement la question du pourquoi,c'est pas suffisant non plus. Nous sommes dans un monde beaucoup plus complexe. Ce n'est pas le 18ème,ce n'est pas le 19ème siècle. Nous sommes à la fin du 20e siècle, et cette complexité ne peut être traduite que si l'on se pose plusieurs types de question en même temps. Ce n'est pas une question de valeur.
Je vais résumer Pourquoi et Comment avec un dessin. (fig. 3) C'est une histoire. Cela c'est un site,un terrain. Alors,sur ce terrain,il y a quelqu'un qui est là et quelqu'un qui est là. Ca ce sont deux hommes. Ils vivent heureux. Ils sont heureux. Arrive l'architecture. L'architecture,c'est construire un mur. L'architecte est arrivé. Et donc là il y a un mur et ce mur,c'est la séparation.
Donc,la première question,est pourquoi je construis un mur. Car lorsque je construis un mur,je sais que je vais interdire le passage d'un

165

从学院派建筑中解脱出来的一代

学生——葛兰巴克或者努维尔先生等人都是佩罗先生的上一代人,他们的学生时代是一个动荡的时代,是美术学院变革的时代,和他们这一代人相比,您们这一代人让人明显感到对于职业技能训练或者考试制度没有多大的热心,这个差别是从什么地方产生的呢?换句话说,对于佩罗先生您们这一代人来说,您们是怎样看待美术学院的象征——学院派建筑思想的呢?

佩罗——我认为葛兰巴克或者努维尔先生他们那一代人在很大程度上还是受到了美术学院思想的影响[1],而我们这一代人具有更加开放的意识和自由的精神。

学生——这种自由来自哪方面的解放呢?

佩罗——学院派这方面。我认为让·努维尔先生他们这些人打破了学院派思想。所谓的学院派,就是重视形态等形式上的东西,或者说对这些东西感兴趣。新一代的建筑师对形式之类的东西没有兴趣,他们对自然,或者说广义上的自然,以及事物之间的关系等问题感兴趣,不会陷于对形式的追求。从这一点上来讲,

[1] 很大程度上还是受到了美术学院思想的影响:参照本书中努维尔先生关于接受美术学院教育时的讲话。

côté à l'autre. Donc ça c'est la question éthique. Pourquoi je sépare?
Et ensuite la question est comment je sépare. Est-ce que je vais séparer en construisant un mur en pierre?Est-ce que je vais séparer en construisant un mur en bois?en verre?en métal? transparent?opaque?translucide?Comment je vais séparer?
Et c'est pour ça que ces deux questions sont inséparables. Pourquoi et comment:le concept et la matière.
ÉTUDIANT:
Qulles sont des différences entre vos générations et celles précédentes?

DOMINIQUE PERRAULT:
Oui. Bien sûr,il y a des différences. Mais la plus grande différence c'est peut-être que ces architectes qui ont été cités ont connu ce que nous on appelle l'ancienne école, c'est-à-dire qu'ils ont connu les Beaux-Arts, et qu'ils ont été influencés par une école encore académique.
Alors que ma génération a une chance extraordinaire,puisqu'elle arrive après:elle est beaucoup plus libre d'esprit d'une certaine façon. On le voit,ma génération et encore celle qui est plus jeune,sont beaucoup plus ouvertes vers une architecture qui n'est pas académique,

学院派建筑思想对形式非常执著。

大野——让·努维尔先生也是学院派吗？

佩罗——可以说让·努维尔先生本人是打破学院派思想的人士之一，打破的意思也就是说他还是处在那个时代，因此，他仍然是前一个时代的人，与我们不同。这就是人生。（笑）

大野——非常感谢。（鼓掌）

佩罗——谢谢。（鼓掌）

欧盟法院扩建工程
(European Court of Justice Extersion) Photographer Perrault Projects

qui est beaucoup plus en relation avec la culture contemporaine.

ÉTUDIANT:
Des générations précédentes,par exemple Jean Nouvel,elles étaient sous le règne du académisme?

DOMINIQUE PERRAULT:
Non ce n'est pas la même chose. Non,le premier architecte français qui rompt avec cette approche académique,c'est Jean Nouvel.
C'est lui qui crée la rupture.
Mais quand je parle d'académisme,je veux dire que c'est une approche de l'architecture qui s'intéresse à la forme. L'académisme,pour moi, c'est la question stylistique. C'est la question de la forme.

Pour ma génération et les générations qui suivent, la forme ne nous intéresse pas. Le design ne nous intéresse pas. Ce qui nous intéresse,ce sont les questions d'émotion directe,d'aléatoire,de hasard, de nature,de relation avec des dimensions qui sont des dimensions qu'on ne peut pas définir par les formes. Des formes indéterminées,voilà.
Cette rupture,c'est grâce à Jean Nouvel. C'est certainement l'architecte qui a donné du sens à cette rupture,mais lui historiquement fait partie d'avant, comme nous on fait partie d'après,puisque nous, nous sommes après la coupure. C'est la vie. Aligato.

座谈会

建 筑 与 教 育

——如何制造"场所"
1999年1月28日

千叶——这个系列讲座的策划，是在安藤忠雄老师的倡议下开始的。安藤老师有两个愿望，其一是邀请风格各异的建筑大师来这里，不是展示他们的作品，而是向我们讲述他们年轻时期以及学生时期的学习和生活经历，希望这些经历能够对将来从事建筑工作以及以建筑师为目标的学子们有所启发和帮助。安藤老师的另外一个愿望，就是希望学生们在聆听学习什么、怎样学习的同时，与当事人近距离接触，直接感受现场的气氛。

今天，我们召集当时对建筑大师们进行访谈的老师进行座谈，各位现在同时也在大学执教，首先希望大家能够讲一下当时采访的感想，以及留下的印象。

大野——我现在就正在努力成为一名建筑师，因此，这次的讲座对我来说非常有意义。能够直接对这些大师进行访谈，实在是太好了。谈到这个系列讲座，我个人的情况是，到采访佩罗先生的时候，我对提问的

座谈会情景

方式以及现场气氛都有一些感觉了，也知道该怎样提问了，还算好，可最初对皮亚诺先生进行采访的时候，因为是第一个，我自己也非常紧张。他一开始就谈到自己根本就没有进过大学，我一下子就不知道该怎样往下问了，非常尴尬……（笑）

我在大学里执教，同时也进行建筑设计，自己开始做设计时❶就已经在大学里了，因此对佩罗的讲话印象非常深刻。而且在这次采访之前我刚刚去过"法国国家图书馆"，有很多感想，正想就此和他进行交谈的时候他刚好来了，从这个方面上来说时间也非常巧。

他给我们讲了他以硕士论文的形式研究了巴黎的区政厅是怎样在19世纪成立的，之后又在国立土木大学进行了关于修道院的研究，他将他对修道院中庭和巴黎城市结构的研究成果非常有效地运用在图书馆的设计之中。对建筑师来讲，怎样进行构思是非常重要的事情，能够从他那里听到关于这些构思的具体讲解真是一件有趣的事情。

另外，我对这两人还有一个非常深刻的印象，那就是如果不是非常热爱建筑的人是不可能成为建筑师的，这一点又一次从他们身上得到了确认。听到两个人都讲到他们从年轻的时候就开始从早到晚沉浸在建筑的世界里，我立即重新对自己进行了一次反省，看来，从现在开始必须要更加努力才行。

岸田——皮亚诺、努维尔、雷可瑞塔、盖里、贝聿铭、佩罗等人，都是当今具有代表性的建筑师，他们能够建造出"大写的建筑"❷。这次讲座让我们连续见到这些建筑大师，无论是对我还是对学生来讲都是一次非常好的策划。

我负责采访的努维尔和贝聿铭两位先生，不仅在年龄上，在设计思路上也是非常具有对比意义的，这是一件意味深长的事。他们虽然是风格和思想完全不

❶ 自己开始做设计时：大野秀敏教授与中野恒明先生一起从1984年开始主持阿儒普综合计划事务所并进行设计活动，当时他已经在东京大学工学部建筑学科桢文彦研究室做助手。

❷ 大写的建筑：原义是勒·柯布西埃在《走向新建筑》一书中主张的具备从古典建筑中一直延续下来的不变的建筑本质的建筑。矶崎新在"筑波中心"完成的时候，在1983年11月的《新建筑》杂志上发表了论文《探讨都市、国家以及式样》，文中也使用了这个词汇，意思相同。

同的建筑师，但超越年龄的隔阂，我也强烈地感觉到他们中间有着共同的东西。同时，我也意识到，如果没有这样的基本素质，他们也不可能设计出"大写的建筑"，而最终成为"建筑大师"。

另一个非常深的印象是，他们能够最终到达"大写的建筑·建筑大师"这样的境界，首先是源于他们本人的人品，也就是人格的魅力，这一点起到了非常重要的作用。虽然这个问题我可能是在这次见面之后才认识到的，但仍然是一件非常好的事。

千叶——能不能具体讲一下。

岸田——首先，是不拘泥于学习一种知识。他们从一开始就兴趣广泛，广泛地吸纳各种知识，才逐步成就了他们的现在，我有这样的印象。

铃木——非常感谢能有这样一个有趣的机会。我是初次和雷可瑞塔先生见面，对我来说，这是一个非常好的经历，非常感谢。

一般来说，建筑师讲话的时候，往往会以最近的工作为话题，而"我是怎样成为建筑师的"之类的话题就别有一番意义。我本人也是这样，我曾经就"卢浮宫美术馆"的玻璃金字塔对贝聿铭先生进行过采访，不自觉地就问到了对现在正在进行的项目的思考，以及怎样将它实现等等问题。在关西机场建成时我对皮亚诺进行过采访，也是通过询问最近的工程情况来了解这位建筑师的。而这个系列讲座则不同，它的目的在于探寻一个普通人怎样成为建筑师。这次讲座中的任何一个建筑师的故事都非常有趣。对学生来讲，这或许比单纯地谈建筑更能"真实"地、近距离地看懂他们。

正像岸田先生所说的那样，六位中任何一位都兴趣广泛、重视旅行，这一点是他们的共同之处。我想，能够成为一位建筑大师与这有着密切的关系。这次讲

❶ 人们的欣赏趣味非常有局限性：例如，以透明性为主题建造建筑的建筑师们会形成一个流派，现在这个流派可能被认为是主流，其他的就会被认为是次流。这句话是针对日本这种风潮的批判。

❷ 与阿尔瓦罗·西扎先生一起去看了日本的寺庙建筑。弗兰克·盖里先生这次来日本，是为了参加高松宫殿下纪念世界文化奖的十周年纪念典礼。由于是十周年纪念典礼，因此邀请了历年的全部获奖者。阿尔瓦罗·西扎是本年度（1998年）的获奖者。历年的获奖者中，还有贝聿铭，詹姆斯·斯特林，盖·奥连提（Gae Aulenti），丹下健三，查尔斯·柯里亚（Charles M.Correa），皮亚诺，安藤忠雄，理查德·迈耶。

岸田省吾

座让我们感受到每一位谈话者的魅力的同时，也让我们认识到这是建筑大师的核心内容之一，这应该是一个非常大的收获。

梵德雷——我以前听过盖里先生的讲话，知道他是一位谦虚的人，因此，这次我就希望他能够直率地回忆自己，结果也正是这样。

盖里先生与安藤先生进行了友好的交谈。虽然他与安藤先生在建筑手法上相差很远，但却能够互相理解。这应该源于他们对建筑的广泛了解，非常令人感动。对日本人我有这样一种看法：人们的欣赏趣味具有局限性❶，和自己思想趣味相同的人在一起，他们愿意交谈，和其他人在一起就不一样了。

盖里先生在东京大学讲演的第二天，就与阿尔瓦罗·西扎先生一起去看了日本的寺庙建筑❷，我想他们又会在创作其他类型作品的时候，在更深的层次上进行切磋理解建筑。

他的讲话对我触动很大，成长也好，发展也好，他有着非常强烈的完善自己的愿望，而且拥有不断改变自己的能力。同时，这次讲座让我们深切地体会到他年轻的时候也是历经艰辛的，他完全靠自己的努力成为了建筑大师。这一点对于东京大学的学生来讲是有特别意义的。我强烈地感受到，正是因为经历了艰辛，才拥有更多的发言权。他现在虽然已经是70岁了，但仍然不断地创作新的作品，非常有活力。看来我自己也要更加努力才行（笑）。

岸田——与之相对的是法国的建筑大师。受了相当典型的高等教育。

大野——确实是这么回事。

梵德雷——在听到贝聿铭先生的演讲之前，我一直认为他是一帆风顺的，没想到他是乘船从中国来到美国的，当时的事情一点也不知道……

岸田——贝聿铭先生最初是想成为工程师的，但最后却选择了建筑师的道路。一旦认识到另一个世界更适合自己时，立即毫不犹豫地进行改变，这一点真是不简单。

大野——不管怎么说，大家都是全力以赴。这一点让我深受感动。

铃木——对。而且大家都非常敏锐。虽然都是高龄，但仍然很年轻。正因为这样，年龄的增长赋予自己的丰富经验又反过来拓宽了自己。

千叶——他们有这样的一个特点，不管到了什么年龄，仍然相信各种可能性。

作为场所的大学

千叶——他们的讲演中有几个共同的话题：旺盛的好奇心、丰富的经验、爱好旅行，我感到，保持广泛的兴趣这一点是可以通过大学教育培养出来的。

岸田——这恐怕也因人而异吧。在巴黎受过高等教育的人，可能会受到强烈的文化背景的影响。而对盖里先生来说，美国文化中的下位文化成了他的文化背景，而这也正是他接受各种事物的基础。反正情况并不完全一样，经常拓宽自己是他们的共同之处。

大野秀敏

大野——在我采访的三个人当中，只有佩罗先生是最好地利用了大学的。努维尔先生"68年"的时候根本就没怎么去大学，皮亚诺先生也总是在弗兰克·阿尔比尼之处打工，看来，即使不进大学，也可以培养出建筑师，我倒觉得这样也行。

岸田——努维尔先生是在"美术学院"中接触到了很多非常了不起的人，也许是因为这种接触的机会比较多……

大野——20世纪60年代末的大学非常有活力。我想当时如果不在大学，就不可能有这样的经历。对他来讲，与其说是"作为制度的大学"，不如说是"作为场所的大学"。

千叶——听到这样的谈话，就会让人怀疑大学中的建筑教育能否在培育建筑师方面起到应有的作用……

铃木——我想这是不应该怀疑的。也许在大学中做过什么课题、学过什么科目，并不是很重要，但大学给了他一个"场所"。例如皮亚诺先生，他讲到他把大学当作一个睡觉的地方，精力完全投入到设计当中去了，但换句话来讲，正因为有了大学这样一个场所，才能够从那里出去画图，对吧。也就是说，虽然知识更新的速度很快，今天学习了最新的知识，明天就会变成陈旧的东西，但是，那个时候你在那里了，你就有机会进入这样一个领域，能够和一些人见面、和一些人交往、能够接触到一些东西，从整体上来看，最能够提供这种机会的场所，还应该是大学。在大学里有一个关于建筑的"核"，这一点意义重大，我想将来也仍然会有重大的意义。

岸田——有老师把大学称为"相逢的场所"，从这个意义上来说，各位应该是这个场所的最大的受惠者。

凯瑟琳·梵德雷

大野——虽然皮亚诺先生说他只是在夜间回到大学睡觉（笑），但在那里他肯定会有机会和朋友们一起谈话、讨论等等。

铃木——像这次这样每个月或者两个月一次邀请建筑大师，尤其是非常有特性的建筑大师到大学来访问，是一件意义非常深远的事情，我想这或许是安藤老师对大学所应有的作用的思考结果。我感到东京大学变成了一个非常好的场所。

大野——正是这样。

岸田——本来，原始的大学就是设在桥头或者街头广场的一个学习场所,听讲的人按课程缴纳听课费,这次的系列讲座,就好像回到了12世纪大学的原始状态。

铃木——有趣的是,建筑师总是把同行称为同僚。不管是有丰富工作经验的人还是刚刚开始工作的人,就凭都是建筑师这一点,互相之间就会有共同的话题。我想这是一件非常好的事情。因此,即使是学生,听到大名鼎鼎的人的讲话,也会很快吸收并成为自己的营养。

大野——在世界上不管走到哪儿,只要看到图纸大家都能够理解,这就像一种共同语言一样。那个人在思考什么、他能够做到什么水平,立即就能够知道。不管走到哪个国家,就像铃木先生说的那样,都可以确认自己处在同一个圈子里。

千叶——旅行也是一起观察、一起体验的一个很好的机会,同时,也是一个理解不同观点、不同东西的最好的机会。

岸田——"旅行"就像一个关键词一样,好几个建筑师都把它提了出来。雷可瑞塔先生所说的旅行是一种与字面意思相同的旅行,但是,如果再认真地想一下,对于佩罗先生和努维尔先生来说,他们应该说是花了大量时间进行了另一种形式的旅行,即与不同行业的人相接触,了解不同的世界。

铃木博之

大野——佩罗先生虽然说不旅行,但这个讲演结束之后,说了一声要去看妹岛的建筑作品,转身就去了。

千叶——他说过喜欢一个人在街上走。

岸田——他换了许多学校。我想这种改变场所的做法也应该可以说是一种"旅行"。

铃木——如果读过19世纪建筑师的传记,就会发现大家认为与建筑师一起旅行是最糟糕不过的事情。说实话,我现在也还这么认为(笑)。在旅行的时候,如果没有发现什么有趣的东西,建筑师就站在那儿不动,而且，当你拿出相机站在前面露出微笑正准备拍照的时候，他又会说你"碍事，走开"（笑）。当然，建筑界同行的人还是可以一起去的。

大野——不错。拿着相机镜头朝上拍的人，肯定是建筑师。（笑）

建筑师的背景

大野——在努维尔先生的讲话中，他提到那个建于中世纪的家，他在那里出生和成长。那段讲述非常细腻而且有诗意，让我印象很深。从他的建筑中，我们能够看出他非常关心"表面"，这一点从他这种独特的讲话方式中已经能够看出。真是非常好的讲话。

铃木——他生长在一个小镇的17世纪的贵族宅邸中，从这一点可以大致推测出他的阶级背景，也就是较为富裕的中产阶级的知识分子家庭。他是一个感性、敏锐的人，能够鲜明地记住自己的成长环境，并影响到现在的建筑设计。虽然说对一件事情太了解了也就没意思了（笑），但是，了解他的这种典型的法国知识分子文化背景是非常有意义的。

岸田——但是，努维尔先生的情况还不能仅从背景来说明，我觉得他的性格中有非常细腻的一面。生下来就是这样。

铃木——是这样。巴黎卡梯尔基金会画廊，本来是夏特布里安❶居住过的地方。建筑用地就在当时的废墟之上，在建筑的形象设计中，以前的旧建筑空间被设计成一种透明的要素，在新的空间中得以体现。这种传统的空间构成源自他自身的原始体验。我个人认为在努维尔的作品中，巴黎卡梯尔基金会画廊是至今为止

❶ 夏特布里安（Chateaubriand）：法国浪漫主义文学大师。全名是法兰索瓦·莱·德·夏特布里安（Francois Rene Chateaubriand）。曾居住在卡梯尔基金会的用地内。建筑用地内仍保留了他种植的杉树。

最好的作品，从他的谈话中我们联想到他的文化背景和这种建筑的关系，听的时候感到意味非常深远。

千叶——在皮亚诺先生身上也可以看到现在所讲的这种原始背景的东西。他受到他的家庭——建筑世家的强烈影响。他的设计中有许多手工的特点，他往往一边用各种部件和材料进行大量的试做一边进行设计。

铃木——我觉得他的做法非常有意大利人的特点。他肯定比较喜欢材料质感之类的东西。虽然他的建筑具有高技派形象，但他也常用瓷砖之类的材料，也会设计像木桶一样的建筑。喜欢金属感觉的人突然用起了瓷砖或木头等要素，往往容易让人搞不清楚是怎么回事，但他并不是这样。

大野——这也许就是工匠的思路吧。我从小就非常喜欢做模型，因此也非常能够理解他。也就是说不管是金属还是木头，重要的是组合时的快感，他沉浸在这里面……

铃木——除了日本，在世界上，这一点也就是意大利残存得比较深吧。

岸田——我认为皮亚诺先生更有类似于英国工匠的精神。

梵德雷——我看他的建筑的时候，会感到他的建筑是"活的"。他往往在开发具有工艺性质的东西的时候，也给它提供了一个发展和变化的可能性，因此，他应该是关心各种材料的。所谓"发展"，也就是说在经历了时间、重叠了历史之后，事物会渐渐地发生变化，他关心这种变化。努维尔先生在巴黎卡梯尔基金会画廊这栋建筑中也非常重视利用旧的东西，在此基础上加入新的东西。我从日本出发去看这栋建筑的时候，强烈地意识到自己和他们一样是欧洲人。这表现在重视并认真思考某种东西所拥有的历史，并在这个过程中思考新的事物这样一种思考方法上。现在这个时代重视"变化"。从这个意义上来讲，皮亚诺先生的设计有手工制作的部分，重视材料这一点也表示他有"变化"的意识。

铃木——所谓能够进行"变化"……

梵德雷——他现在好像对南太平洋非常关心，我想这是因为那里是最符合这种条件的地方，风的流动、太阳等等，我想他是想抓住最能反映变化的东西。

AA学校的教育与东京大学的教育

千叶——还是回到教育的话题。各种文化背景，以及小时候生活环境的熏陶等等因素都对一个建筑师具有很大的影响，那么，大家认为大学教育应该承担什么部分好呢？

岸田——刚才铃木先生也说过了，我想，大学作为一个"场所"是非常重要的。我不清楚能不能这样说，美国的职业学校❶是进行彻底的职业教育的地方，这种教育方法总是会有一个界限的。因此，我感觉到大学的存在理由，更在于能够提供什么样的"场所"。

千叶——凯瑟琳女士是AA学校❷毕业的，AA学校一开始也正是像沙龙一样，是一个各种人交往的场所，对吧？

梵德雷——是这样。与日本的教育非常不一样的是，日本的教育重视最终成果，而AA学校是针对每个人的情况一步步地进行教育，对结果则不那么重视。那里非常重视教育学生能够自己发现各种事情。

岸田——这让人深刻地感受到国家的不同。但是，不管在哪里，职业学校进行的教育是非常必要的，就这一点来讲，不管是哪个国家，都必须学习基本的技能。例如，要成为建筑史专家的人就必须学习阅读历史

❶ 职业学校：日本的大学把培养学者当作主要目标，与此相对，美国的许多大学则致力于培养技术人才。商业学校、法律学校等就是其典型实例。

❷ AA学校：正式名称是Architectural Association School of Architectural，1847年创立，历史悠久，现在是欧洲最有名的建筑学校之一。起源是被英国皇家建筑学会（RIBA）拒绝的年轻设计师们为了互相讨论而成立的会员制俱乐部。该校采用联合教学的独特的教育方法，培养了众多的有特色的人才，不仅在建筑界，在其他专业领域也人才辈出，因此享誉世界。这种AA风格的形成，与阿鲁宾·伯阳斯基校长密切相关，他从1971年开始任校长，直到1990年突然去世为止。该校的毕业生中有许多著名人物，如这次没能来日的雷姆·库哈斯，理查德·罗杰斯，库伯·西迈布劳（Coop Himmelblau），大卫·齐伯费尔德（David Chipperfield）等，活跃在日本的有汤姆·海奈甘（Tom Heneghan）和凯瑟琳·梵德雷等。

资料的方法，而对于建筑师来讲也有同样的部分。

我认为AA学校与东京大学的反差在于着重点的不同，日本可以说是比较重视"形"，因此也经常被称为是"形的文化"，这可以说是建筑师的基本技术，建筑师必须在学习了这种基本知识之后才会逐渐变得熟练，等到成名之后就可以打破"形"，或者说走向不同的"形"。从这种解释来看，我想两者是有所不同的。

大野——东京大学以前的教育是这样一个流程：二年级的第一个学期学习透视图的画法，或者把有名的建筑师的图纸拿来进行描图❸，进入三年级之后才能进行设计制图。那时经常给学生举的一个例子，就是在学打网球的时候，必须先空挥几次拍子，在有形之前是不能进入场地的。就是这样的一种教育方法。我想，如果一开始就非常了解整体情况，那么这种教育方法是有效的，就好比打网球一样，比赛形式非常明确而且已经格式化，在这样的情况下是可以的。但是，建筑师的专业能力是不停地变化的，他无法看到最后的状况，也就不清楚学习的目的，在这种情况下只是告诉学生画线的话，学生不可能明白这种训练的意义。因此，二年级的设计制图的教育方法现在也逐渐地发生了变化。

针对这种情况，三年前我们开始了工作室制度❹。以前的建筑教育是基于建筑计划学这样一个大的教育体系的，其教育方法就是将建筑以类型划分，从简单的类型开始，分阶段、分设施进行学习。但是，建筑并不是这样一种单纯的程式化的东西，它有着各种各样的切入方法。建筑的学习往往会是这样：有人想做这事，有人想做那事，老师也一样，有的老师关心这个方面，有的老师关心另一个方面。工作室制度就是希望能够把这种个性和多样性引入设计教学中，避免一年中都学习同样的东西。这个制度现在已经渐渐地

❸ 把有名的建筑师的图纸拿来进行描图：在东京大学，专业教育是从二年级的下学期开始的，在最初的半年中，设计制图的课题主要是对柯布西耶、赖特、吉村顺三等人的图纸进行描图和绘制透视图。这门课曾因为题目数量太多，时间过紧，学生负担过重而出名。近来由于设计课开始的时间提前了，相应减轻了学生的负担。

❹ 工作室制度：设计制图课程的授课方法之一，就是为了向学生传达设计是可以从多种视角切入的。在通常由设计老师担任的设计制图课程中，加入结构和历史课程的老师，每个老师提出自己的课题，学生根据自己的兴趣进行选择，最后进行集中讲评。照片是讲评会的情形。

181

固定了下来。

岸田——我并不是说以"形的文化"为基础的教育是好的，相反，我感到原因可能也正是出在这里。在引入了多样化思想，价值观也发生了变化之后，教育体系也必然会发生变化。但是，工作室的课题也好，现在进行的改革也好，就拿造型基础❶这个新的课程来讲，在这个思考框架中，一定的系统知识和基础技能等这些东西仍然是存在的，因此，我想，以"形的文化"为基础的教育也仍然是有好的一面的。

梵德雷——但是技能如果比构思先形成的话，构思就会变得迟钝……

岸田——我想并不是谁先谁后的问题，两者应该是同时的。

梵德雷——对，可以这样说。

铃木——在东京大学，进入二年级的后半学期才终于开始学习建筑。从这一点上讲，与那些一入学就进入建筑学科的学校相比，或许会被人说晚了，但是建筑的学习呢，还是应该至少先学习一些非建筑的东西之后再开始。在美国也是这样，在大学中先学习的全是些与建筑无关的东西，在这之后再决定是否学习建筑。我感到高中毕业之后不要立即就学建筑，一个"游荡"的时间是非常重要的。

岸田——这一点我深有同感。现在还保留着教养学部这种体制的大学也就只有东京大学了。甚至还有评论说两年的时间太短，对于建筑学习来说，或许应该把大学教育作为基础教育的一环来进行。

千叶——AA学校的学生中，一开始就以建筑师为目标的人多吗？

梵德雷——最初是有很多人想成为建筑师的，但渐渐地在学习的过程中，有转向电影专业的人，也有对经商感兴趣并开始经商的人，而且还不少，还有转向电

---

❶ 造型基础：为了扩展建筑学学生在造型设计方面的视野而开设的一门课，进行绘画，雕刻以及照明，家具设计等练习。现在画家木下晋，照明设计师内出薰，家具设计师藤江和子等人任兼职教师，讲授这门课。

视采访的人，总之，有相当广泛的范围。成为职业建筑师的意识在 AA 学校并不是那么强烈。

岸田——AA 学校出了很多个性丰富的人，这一点非常出名。

梵德雷——库哈斯❷是学了电影之后转过来的，哈迪德❸(Zaha Hadid) 是学了数学之后转过来的。学习了别的东西之后转过来的人可以说更有潜能。里布斯金❹不也是学过音乐的吗，他还学了数学和哲学。

千叶——AA 学校没有诸如结构、历史等这样的固定课程。学习的方法真是非常特殊。

梵德雷——是。根本不会画透视图、不会结构计算、也不怎么了解历史的人也有。与此相反，比结构老师和历史老师都能更有意思地进行讲解的人也有，我想正是这使它成为"前沿"。

大野——那么技能被放在什么位置了呢？刚才你在讲话中提到，在教育中如果太早地教给技能就会影响构思，但是，如果没有什么技能的话就无法进行表现呀，尤其是对这种进行视觉方面学习的人。

梵德雷——这个嘛，周围的人都很熟练，受熟练的人的影响也就自然地会画图了。因此，在图纸的画法上也会有流行的倾向，在一定的时期中大家都画类似的东西。但是有一天，自然地就有谁发现了其他的画法并不这样做了，这时老师也不会强制说一定要怎样画。

千叶——像这次演讲的大师一样，通过在事务所打工学习技能的人多吗？

梵德雷——也有这样做的人，主要是没有钱的人。AA 学校学费非常高，因此会有一边工作一边学习的人。但是老师们并不认为这种事情多么重要。

❷ 雷姆·库哈斯 (Rem Koolhaas)：1944 年生于荷兰的建筑师。这次在临开讲之前取消了讲座。他的激进的著作和作品现在仍然不断地影响着世界上大量的建筑师。他在成为建筑师之前曾经做过电影和戏剧的编辑，这成为他创作《疯狂纽约》的文化背景。他的这段历史非常有名。

❸ 哈迪德 (Zaha Hadid)：1950 年生于伊拉克巴格达的建筑师。在 1983 年香港高峰俱乐部 (The Reak Leisure Club) 设计竞赛中一举成名。他受到构成主义的强烈影响，以歪斜的设计著名，这是第一次听到他曾学过数学。

❹ 里布斯金 (Daniel Libeskind)：1946 年生于波兰的建筑师。以概念表现和让人想起锐利金属的建筑表现著称。在纽约和以色列学习音乐，在纽约的库伯联盟 (Cooper Union) 学习建筑，在英国埃塞克斯 (Essex) 大学学习历史和哲学。

培养自由的创造力

梵德雷——我在AA教书的时候，有人立即就能做出很好的模型，也有人做不好，但是，做出最有趣的模型的人是那些做不好模型的人（笑）。能够很快地做出好模型的人把精力集中在做上面，构思并不自由，技能变成了负担。结果最发愁的正是这些人。模型虽然做得不老练但想法很有趣的人，得到模型做得好的人的帮助之后成为了第一名（笑）。因此，我想不要一开始就定下各种条条框框，做出好的东西就行。

铃木——是这样。教师还是不要说一定要怎样做之类的话好。

大野——这些话可以用在历史的学习中。

铃木——我也这样想。最初找到什么课题会决定以后做什么。在方法上有以时代来决定的方法，也有以地方来决定的方法，还有以类型来决定的方法，这时，他人不该诱导他说文艺复兴方面的东西比较有趣，或者现在是技术史的时代等等。某个人对什么感兴趣，从最初到最后都是"这个人"的问题。

大野——这方面的事情已经开始引起思考了。不过，学校可以提供一个场所，使他知道怎样做才能找到正确的方向，关于这个问题大家是怎么想的呢？

铃木——学生说想要做这样一个课题时，如果能给他一些建议，例如对他讲这样的东西读一下怎么样、看一下怎么样、去问一下这些人怎么样等等，不是很好吗。

大野——当一个引导者。

铃木——嗯。这样教师就处在被询问的位置上。如果有学生对不懂的事情感兴趣的话，可以试着告诉他：

虽然你希望这么做,但最好的道路可能会在另一处吧。实际上学生试着走一下,又说对那里不感兴趣,这时再告诉他:那么,这样怎么样?我想是不是可以这样引导呢。从这方面来讲,在大学里学习和跟着建筑师学习有所不同。如果去了勒·柯布西埃那里就会受到他的直接影响,而去了安藤先生那里又会受到安藤先生的影响。但是,大学本来就是一个比较大的场所,这里面可能会出现第二个安藤、第二个勒·柯布西耶、第二个密斯,因此,我觉得应该在某种程度上让它成为一个开放的场所。东京大学有一些个性很强的老师,他们应该会有一些自己独特的教育方法,也许,情况又不完全是这样。

岸田——正是这样的老师视野非常广。安藤老师就是这样一个典型。

铃木——恐怕是这样。

岸田——为了成为这样的引导者,教师自身应该受什么样的教育也非常重要,在建筑专业教育之前的那一部分也很重要。

铃木——这怎么讲?

岸田——矶崎新先生住在学生宿舍里的时候,据说山田洋次先生住在同室。虽然那时的学生宿舍,现在已成为要拆除的对象,但在当时也是大学提供的一个交流场所。在那里可以见到各种各样的人,有不同年级的同学。宿舍也是一个各种思维交流和融合的场所。

铃木——确实,这一点非常重要。如果讲起以前大学宿舍或者过去高中时代的事情,总会有一些流传下来的佳话,虽然这有对过去经历进行美化的成分,但是,不同专业的人住在一起,就会创造出一种交往。现在这种交流的场所逐渐消失了,高中教育本身也在某种意义上变得均质化,教育范围在变窄,即使进入东京大学,接触面也基本固定,建筑师的圈子也是一样的

❶ 尼古拉斯·尼格罗彭特（Nicholas Negroponte）：利用电子技术探索数字媒体和通信的可能性这一研究领域的先驱者。生于希腊的美国人。原本学习建筑，1966年25岁的时候任教麻省理工学院，成为该校年轻的教授。就任后创办了媒体实验室。现在仍在那里任所长，进行着独特的研究活动。他的研究受到各国企业的关注。

❷ 伊藤隆久：1974年毕业于东京大学工学部建筑学专业。1978年修完东京大学研究生院课程。之后进入日建设计，参与了"NS大厦"、"圣路加医院"等许多重要的工程。现在主持加藤隆久都市建筑事务所。伊藤先生作为设计者进行的这项老挝的小学校建设项目，是日本民间交流中心作为教育援助计划的一环从1996年开始的一项计划。到1999年6月为止已经有3座学校竣工。现在该项计划的项目仍在进行中，今后也会继续下去。

❸ NPO：Non Profit Organization的简称，一种进行公益活动的非盈利团体。日本在阪神大地震时也曾接受了公益团体的援助活动。1988年国会还专门通过了NPO法案。

情况。

梵德雷——麻省理工学院的尼古拉斯·尼格罗彭特❶（Nicholas Negroponte）先生有一次在回答"美国人的想法为什么特别丰富"这个提问时说过：这是因为美国有不同国家、不同文化背景的人，就算你想简单地处理问题，也不得不和不同文化背景的人打交道，因此必须变得灵活。日本是一个岛国，没有这种必要，不同文化之间的交流自然就比较少，能够弥补这一点的，是和不同专业的人在一起。现在，东京大学的学生计划在明年的五月祭中搞一个展示场，希望能够吸引音乐、诗歌、演艺界的人到这里来。我感到他们十分想要和外界接触，得到外界的刺激。应该支持他们。

千叶——从这个意义上来讲盖里先生的讲话给我留下非常深的印象。从他的讲话中我们可以看出他与不同专业、不同工种的人有着很深的交往，各类艺术家的名字也能一个接一个地讲出来。

铃木——我感觉到和建筑师相比，其他专业的人给他提供了更多的支持和帮助。

千叶——安藤先生经常讲大家应该多到海外去，也是同样的意思。这种事情有没有可能纳入学校的教育体系？

岸田——在欧洲有些学校的毕业条件之一就是必须进行一年或半年的实习。这一点非常好。以前，有学生在巴黎工作半年，然后去柏林，再去洛杉矶，这样巡回一圈也就是"旅行"了，也开阔了视野。

铃木——在明治时期工部大学校的时候，学生有一年的现场实习。我想大概是在帝国大学的时候，与其他学院进行了合并，这种现场主义也随之没有了。

大野——去年，伊藤隆久❷先生在老挝做了一个NPO❸性质的学校项目。他进行设计，当地居民和从日本去的志愿者一起进行建造。当时在东京大学招募志愿者时大约有十个人报名，最后好像去了九个。如果以学校为单位来计算，东京大学是去的最多的一个学校。看来如果有机会的话学生们还是希望能够拥有各种经历的。

梵德雷——提到走向世界，日本人就会更多地想到去看西方和西方的建筑。其实现在亚洲也很活跃。我想现在也很有必要回过头来看一下亚洲。在日本，对亚洲建筑关心的人和杂志还比较少。

铃木——是的。不过也在慢慢增加。

对今后东京大学建筑教育的期待

千叶——最后，希望大家能谈谈对今后东京大学教育的意见。

铃木——本来，所谓大学，应该是有各种课程以外的演讲会、工作室、讨论会的。告示栏中应该贴有各种活动的广告。这次建筑大师系列座谈就是这样，虽然来者名气有点过大，但是，卓越的建筑大师一个接一个地来，在促进学科交流的同时，也在某种意义上使学生们产生了一种一体感，共同经历了同一件事情，这种共同的体验对学生来说非常重要。同时，像梵德雷老师所说的那样，这种交流还应该向着更广的文化圈和不同流派的建筑师扩展。不同的东西相互碰撞交流才能形成一个有活力的场所。

大野——我赞成铃木先生的意见。如果可以对其他问题发表意见的话，我认为虽然东京大学一年半的基础课程教育非常好，但从整体上来看课程划分得太细，学生能够静下心来思考问题的时间太少，如果让他们能就一个问题主动地长时间地进行思考就好了，比如设计课题，能否考虑一个课题作一个学期？正好现在我们建筑学院有许多做实际工程的老师，这些实际经验会帮助学生更细致、更有深度地进行学习。应该根据不同情况对课题的设置进行改变，也应该让学生有多种途径进行思考，给学生一个能够静下心来投入进去的机会。

还有一点，可能无法立即实现，就是如果能像其他很多建筑学校那样，有一定的作业空间、工作室❶就好了。不管怎么说，建筑是一种和具体物体打交道的工作，物体是怎样形成的，必须一边制作一边思考才能体验出来，这一点非常重要。所以说，要是能有这样的工作室就好了。

千叶——看了AA学校的学生作品后我感慨很深，他们一个题目可以花一年的时间来做。在这一年内学生可以首先进行历史调查，然后在某一个时期到现场去调查研究，并在那里制作一些什么，或者在某一个时期做出一个方案或者报告等，学生可以从各种层面上进行研究和设计。用这种方法，即使是能力不强的学生也有机会通过其他形式得到教育。现在东京大学的教育方法适合能力强的学生，能够在短时间内完成课题的学生会得到很好的培养，但其他学生有时就会得不到承认。

岸田——先不说一个课题做一年会怎么样，就是做半年，也明显就会有相当充裕的时间进行学习。现在的本科教育和研究生教育是有明显划分的，欧洲国家现

❶ 工作室：为了制作建筑模型，试作家具、照明灯具等而必须的放置木工工具、机械等物品的作业空间。这种工作室在欧洲和美国的大学中非常普及，是学生能够亲自体验制作过程的一个非常重要的设施。

学生画的素描

在都改成了五年教育，我个人感觉东京大学也必须顺应外界变化，对自身进行更新。

还有，铃木老师刚才讲到的，广泛交流也是一个重要的问题。例如在设计课程中，不仅仅是出题的老师，也让外部的人在最终评图时能够发表意见，这样就会有更多的人参与进来。

梵德雷——我对大野老师的讲话有同感。学生们也不能仅是接受，如果能更主动一些就好了。课题出来之后，首先要自己下功夫去现场进行实地调研，并进行纪录。带着相机、拿着速写本，进行录音、拍摄等工作。首先通过这些方法获取第一手资料。然后在下一个阶段，根据这些调查进行设计。我认为必须这样提高学生的参与意识。

例如，在建筑与城市的关系这一问题上，AA学校就非常重视学生的直接参与意识。这样即使将来走出大学参加工作以后，学生们也会有意识努力让环境变得更好。在这个课题中，必须注意对环境的分析不能只采用图表分析之类的方法，这种方法与现实有一定的距离。学生们必须捕捉与身体有关的感官体验，这一点非常重要。在这里，使用自己的身体进行调查和思考这一点非常重要。在大学里，如果学校对此有热情，学生们也会对此做出积极的反应并变得更加努力。在日本，听到"激情"一词，大家都会露出不以为然的表情，但是我认为，对建筑的激情、对环境的激情，与重视建筑的视觉形象一样重要。

岸田——日本人也不是说没有激情，可能是表现的方法不同吧。

梵德雷——当然，激情还是有的。前一段时间在二年级的设计制图课程中，我出了一个题目，是让大家绘制一个神圣的空间。课题一出来，大家都感到非常困惑。我把建筑空间的照片复印之后交给他们，让他们用炭笔在大纸上画。最初大家都是对照片进行忠实的描绘,但渐渐地就开始用自己的眼光来看,来思考,并开始对其他的对象进行描绘了。最后气氛变得非常热烈,也产生了群体意识,大家都很兴奋,有了激情。从他们的表情中可以看出,他们找到了属于自己的东西。

千叶——这种方法非常有效地引发了大家的激情,他们因此而感动。

铃木——其他人看起来,最初会有感到奇怪:在做什么呢?大家好像都在画同样的东西……但是,渐渐地学生们变得很有生气,这一点非常有意思。

梵德雷——大家常说帕提农神庙是非常好的建筑,或者勒·柯布西埃是怎样怎样做的,等等。但是学生们很难实际进入那个空间进行体验,这时候怎么办?研究它的图纸,自己也进行绘制,在这个过程中,就会一点点地体会到那个空间的感动,并理解它的本质。在这之后,就能够用新的眼光去看待其他的事物了。在这个课题中也有学生选择了东京大学的建筑,我想他一定发现了在自己普通生活中也有好的空间。不一定非得到意大利去看好的建筑,各种地方都会有高水平的建筑。普通的街道也有好的地方,木造的空间也很美丽。当学生们具有了这些能力之后,下一步的任务就是怎么把这些带入设计之中。

大野——我想这里有一个教育技巧的问题。你是把照片交给他们让他们照着画的。如果一开始什么都没有,只是让他们画的话,他们没有绘图技法,恐怕只能画出一些奇怪的图来。照片本身就已经有了构图,他们可以巧妙地利用这些,并在此基础上发展。我想这作为一种教育方法是非常有意思的。

岸田——激发出学生们一直隐藏着的激情,也是凯瑟琳老师来后东京大学变得更加开放的一个很好的例子。

大野——亲自进行一次绘制,这样的体验是很重要的,我想今后是有用的。

梵德雷——用炭笔擦着画，手也弄脏了，这都是用身体进行的体验。

千叶——用自己的身体进行实际的劳动，这种感觉非常重要。从这个意义上来说，刚才说的工作室的事情也非常重要。

岸田——以前我去法国的美术学院的时候，看到在地下室里有家具工作室那样的地方。学生们分成小组进行作业，非常好。能感受木材这种材料到底有多大的硬度、在什么样的加工方法下会发生什么样的变化，等等，具有这样的感官体验是非常重要的。欧洲的学校中普遍都有这种工作室吧。

大野——在最近的学生中，有的人不会从木纹的方向判断哪个方向强度高。这是因为与物质接触的机会没有了，身体也就不再具备这种感觉。刚才也说过了，建筑是一个与物体打交道的工作，如何把这一点反映在教学中，在今后可能会变得越来越重要。

千叶——这个系列讲座在明年将会把范围扩大到建筑以外的人，有人提出也可以请一些实际进行制作的人，以及艺术家等等。现在我们所讲的建筑建造方面的事情，如果请他们从别的角度来分析也是很有益处的。

梵德雷——另外，我想邀请一些更年轻的人也会很有意思。这次请的人声名太显赫了……对学生来讲，与他们更接近一点的人也许会对他们产生很好的影响。

千叶——重视实际制作和实际体验，同时在大学里为学生们提供一个交流的场所都是非常重要的事情。今天大家谈了这么长的时间，非常感谢。

编译者简历

座谈会出席者简历

铃木博之

1945年生于东京都。1968年毕业于东京大学工学部建筑学专业。1974年同大学工学部研究科建筑学专业博士课程修满后退学，任同工学部专任讲师。1974—1975年留学伦敦大学Courtauld美术史研究所。1978年任东京大学副教授。1984年取得工学博士学位。1990年开始任东京大学教授，至今。1993年任哈佛大学客座教授。获得每日日本研究特别奖、艺术选奖文部大臣新人奖、Suntory学艺奖、日本建筑学会奖（论文奖）等众多奖项。著作有"建筑的世纪末"（晶文社，1977）、"维多利亚·哥特的崩溃"（中央公论美术出版，1996）、"看得见的城市、看不见的城市"（岩波书店，1997）、"日本现代十都市"（中央公论新社，1999）等。

大野秀敏

1949年生于岐阜县。1972年毕业于东京大学工学部建筑学专业，1975年修完同大学研究生院工学部研究科建筑学专业硕士课程。1976年进入桢综合计画事务所工作。1983年开始任东京大学助手（桢文彦研究室）。1988年任东京大学副教授。1997年取得博士学位（工学）。1999年任东京大学教授。主要作品有"NBK关工园办公大楼"（1993年JIA新人奖、1994年通产省优秀设计奖）、"YKK滑川寮"（1994年度中部建筑奖、1996年度日本建筑学会作品选奖）等。著作有 "忽隐忽现的城市"（鹿岛出版会，合著，1980）、"SD330号特集：香港超级城市"（鹿岛出版会，责任编辑，1992）等。

岸田省吾

1951年生于东京都。1975年毕业于东京大学工学部建筑学专业，1980年同大学研究生院工学部研究科建筑学专业博士课程中途退学，之后先后在冈田新一设计事务所、矶崎新工作室工作，1992年任东京大学副教授，至今。1997年取得博士学位（工学）。

主要作品有"新桥SG大厦"、"东京大学武田先端知大厦"、"东京大学工学部2号馆改建"、"东京大学研究综合博物馆小石川分馆"等。著作有"巴塞罗那——地中海城市的存在证明"（丸善出版，1991）、"大学的空间"（鹿岛出版会，编著，1996）、"东京大学"（东京大学出版会，合著，1998）等。

凯瑟琳·梵德雷

生于英国。1973—1979年就读于AA学校。1980—1982年为东京大学研究生院工学部研究科建筑学专业硕士研究生，并在矶崎新工作室工作。1986年成立牛田1+1梵德雷建筑设计事务所，1988年改组为株式会社牛田1+1梵德雷组合事务所。1998年成立USHIDA·FINDELAY(UK)以及K&T ARCHITECTS。同时，1998年开始任东京大学副教授。1999年任加利福尼亚大学(UCLA)客座教授。主要作品有"TRUSS.WALL.HOUSE"（BBC DESIGN AWARDS特别奖）、"SOFT & HAIRY HOUSE"（1994年日本电子玻璃NEG空间设计竞赛作品部门金奖、1996年度日本建筑学会作品选奖）等。著作有"PARALLEL LAND SCAPES"（TOTO出版，间画廊丛书，1996）、"USHIDA FINDLAY"（2G，GUSTAVO GILI，1998）。

演讲者简历撰写者和注释执笔者简历

岩城和哉——雷佐·皮亚诺

　　1967年生于鹿儿岛县。1991年毕业于东京大学工学部建筑学专业。1993年修完同大学研究生院工学部研究科建筑学专业硕士课程。1996年取得同大学博士学位。同年成为东京大学工学部研究科建筑学专业助手（岸田省吾研究室）。现为东京电机大学理工学部副教授。

山下晶子——让·努维尔

　　1969年生于东京。1992年毕业于东京大学工学部建筑学专业。1994年修完同大学工学部研究科建筑学专业硕士课程。1997年取得同大学博士学位。1998—2000年任东京大学研究生院工学部研究科建筑学专业助手。

驹田刚司——理卡多·雷可瑞塔

　　1965年生于神奈川县。1989年从东京大学工学部建筑学专业毕业后，进入香山工作室环境造型研究所（现香山寿夫建筑研究所）工作。现在是东京大学研究生院工学部研究科建筑学专业助手（岸田省吾）、驹田建筑设计事务所主持人。主要的作品有"西葛西APARTMENTS"、"HOUSE9"等。

本江正茂——弗兰克·盖里

　　1966年生于富山县。1989年毕业于东京大学工学部建筑学专业。1993年在修同大学研究生院工学部研究科建筑学专业博士课程时中途退学。后任同大学助手，从2001年开始任宫城大学事业构想学部设计情报学科空间设计课程专任讲师。

冈由香利——贝聿铭

1966年生于东京都。1990年毕业于东京大学工学部建筑学专业。1991年修完北伦敦工科大学医疗设施计划硕士课程。1995年在取得了东京大学研究生院工学部研究科建筑学专业学分之后退学。后任同大学建筑学专业助手，现在任美国南卡洛里那州立克莱姆森大学建筑学部客座副教授，同时参与同大学护理学部的改建计划。主要作品有"菲律宾塔拉克州母子保健中心"。

坊城俊城——多米尼克·佩罗

1962年生于东京都。1985年毕业于东京大学工学部建筑学专业。1987年修完同大学工学部研究科建筑学专业硕士课程。1988年作为法国政府公费留学生渡法。1992年从巴黎Belleville建筑大学毕业。1995年修完巴黎第四大学美术史学科前期博士课程。1997年取得博士学位。在任东京大学助手、早稻田大学理工总研副教授的同时，也兼任一桥、成城、武藏各大学的兼职讲师，现在是文化厅建造物科调查官。

千叶学——座谈会

1960年生于东京都。1985年毕业于东京大学工学部建筑学专业。1987年修完同大学工学部研究科建筑学专业硕士课程，进入日本设计工作。1993年开始任东京大学工学部研究科建筑学专业助手（安藤忠雄研究室），现在是东京大学研究生院副教授。

译 者 简 历

王静

1985年毕业于南京工学院（现为东南大学）建筑系，1988年获东南大学硕士学位，同年进入河南省建筑设计研究院工作，1991年赴日，先后在寿建筑设计事务所和荣光建设株式会社工作，1998年取得日本国一级注册建筑师资格。

现为东南大学建筑学院副教授。出版了《日本现代空间与材料表现》等著作。

从左往右分别为费移山、王建国、王静。

王建国

博士、博士生导师，教育部"长江学者奖励计划"特聘教授、国家杰出青年科学基金获得者。

1982年毕业于南京工学院建筑系，1989年获东南大学博士学位并留校任教。现任东南大学建筑学院院长，出版了《现代城市设计理论和方法》、《城市设计》和《安藤忠雄》等著作，发表学术论文100余篇。

费移山

1995年起进入东南大学建筑系，相继获得本科学位、硕士学位，其中2002年上半年曾在香港大学建筑系作交换学生。现为东南大学在读博士研究生，研究方向为城市形态和城市交通。曾在《规划师》、《城市规划》等杂志上发表过《田园城市——一个世纪的追求》、《城市设计整体性理论》、《高密度城市形态与城市交通——以香港城市发展为例》等论文。

责 编 后 记

　　历经二十一个月的努力，《建筑师的20岁》终于付梓了。其实每本书在正式出现在书店之前，都有很多鲜为人知的故事，这本书也不例外。从编辑的手中转到读者的手中，这中间有过紧张，有过兴奋，有过期待，有过沮丧，有过喜悦，最终回归到平静。原本想用一种特殊的方式记录下其中的点滴，和读者一起分享，但是真正开始落笔时，却发现千头万绪，无从说起了……

　　首先我要感谢本书的六位主人公，国际建筑大师伦佐·皮亚诺先生及其助理Stefania Canta女士，让·努维尔先生及助理Charlotte Kruk先生，理卡多·雷可瑞塔先生及助手Virginie Vernis de Velasco，弗兰克·盖里先生及助手Keith Mendenhall先生和Laura Stella女士，贝聿铭先生及助手Nancy Robinson女士，多米尼克·佩罗先生及助手Sophie Dauchez女士。他（她）们帮助我解决了本书部分图片的版权问题，同时还热情地提供了一些日文版中没有的新图片，为本书增添了新的亮色。最初我和他（她）们素不相识，又远隔万里，当我以忐忑不安的心情给他（她）们发Email的时候，并没有奢望能得到他（她）们的回复。当一封封回信从美国、法国、墨西哥纷至沓来的时候，我有些喜出望外。感谢建筑大师们对本书的支持。

　　我还要感谢我的法国朋友Emmaneual Py先生和在

加州的同学王刚博士以及同事张彤女士，他们帮助我与六位建筑大师和他们的助手取得了直接联系。

其次，我要感谢我的朋友张亦依女士，她旅日多年，是个独立小说翻译人，为了落实本书的版权，前后花了八个多月的时间和精力。

我还要感谢本书的翻译者东南大学的王建国老师、王静老师和费移山女士。能找到最佳的翻译者不是一件容易的事情，我寻觅许久也未能如愿。最后，在同事邹永华的帮助下，我幸运地结识了王建国老师。他们在整个翻译过程中表现出来的细致、耐心、认真，使这本很有意义、很有趣的书的中文版更具可读性。

我还要感谢我的同事徐晓飞和本书的设计师宁成春老师。在徐晓飞的介绍下，我有幸请到了宁成春老师为本书做设计。因为是双语对照，而且还有不少图片，加上原版书精致的版式设计，都给这本书的设计带来了不小的难度和压力。宁老师最终经过深思熟虑，做出了现在这个朴素、干净的版式。

最后我还要感谢负责本书的终审王仁康老师、校对刘玉霞女士以及为本书顺利出版而付出努力的所有同事们。

对于和我一起经历这本书的所有朋友们和同事们，在此一并表示我真挚的谢意。谢谢大家陪我一起走过这二十一个月，你们给我和这本书都留下了温暖的回忆。

汪亚丁
2005年12月于清华园

Epilogue from Editor

After 21 months hard working, World Architects in Their Twenties eventually sent to the press. Like any book, there are many untold stories behind its appearance in the bookstores, and our book also makes no exception. From the hand of editor to the hand of reader, this book has conveyed my anxiety, excitement, anticipation, depression and joyance, and tranquility at last. I hope to record these special stories using special tone of writing to share with our readers, but when grabbed a pen in my hand, I cannot find a single one to start.

First of all, I want to acknowledge six world architects, Mr. Renzo Piano and his assistant Ms. Stefania Canta, Mr. Jean Nouvel and his assistant Mr. Charlotte Kruk, Mr. Ricardo Legorreta and his assistant Virginie Vernis de Velaso, Mr. Frank O. Gehry and his assistants Mr. Keith Mendenhall and Ms. Laura Stella, Mr. I.M. Pei and his assistant Ms. Nancy Robinson, Mr. Dominique Perrault and his assistant Ms. Sophie Dauchez, for granting permission to use photographs in this book. They even warmly offered me many new photographs that are absent in the original Japanese edition, which added high lights to this book. In the beginning, I didn't expect to get all of their replies when I gingerly wrote to them by email, since we didn't know each other and we are thousands of miles apart. However, when warm replies rained in on all sides, from the United States, France and Mexico, I am really overjoyed. Thank you all for supporting our book. I am also grateful to my friend Mr. Emmaneual Py from France, my classmate Dr. Gang Wang at California as well as my colleague Ms. Tong Zhang. They helped me to get direct contact with these architects and their assistants. Secondly, I want to express my gratitude to Ms. Yiyi Zhang, who has been stayed in Japan for many years as an independent translator, and she has devoted eight months to settle the copyright issue for this book. My thanks also go to translators, Mr. Jianguo, Ms. Jing Wang and Ms. Yishan Fei. It is not easy to find the best translators, and I have searched extensively in vain. It is only with the help of my colleague Mr. Yonghua Zou that I am lucky to know Mr. Wang and his colleagues. Their pains-taking effort and patience have made this instructive and interesting book accessible to our Chinese readers. I also want to acknowledge book designer Mr. Chengchun Ning. It is only with the help of my colleague Mr. Xiaofei Xu that I am lucky to know Mr. Ning. The bilingual and graphic features of this new edition, together with the finely designed original version as a comparison add great difficulty and pressure on the design job. After thorough consideration, Mr. Ning comes up with the present simple and neat layout. The last but not the least, thanks go to Ms. Renkang Wang who is the senior editor , Ms. Yuxia Liu who is in charge of proofing and all the persons who have made great efforts for the publication of this book.

My sincere thanks to all my friends and colleagues. Thank you all to accompany me on this journey, and you have left sweet memory to me and to this book.

Yading Wang

December 2005 at Tsinghua Garden

图书在版编目(CIP)数据

建筑师的20岁／王静，王建国，费移山译. —北京：清华大学出版社，2005.12（2024.5重印）
ISBN 978-7-302-11126-9

Ⅰ.建… Ⅱ.①王…②王…③费… Ⅲ.①建筑设计－研究－世界 ②建筑师－访问记－世界 Ⅳ.TU2

中国版本图书馆 CIP 数据核字（2005）第 054777 号

Original Japanese TITLE：［建築家たちの20代］
编者［東京大学工学部建築学科　安藤忠雄研究室］

Copyright ⓒAndo Tadao Laboratory, Department of Architecture, Graduates School of Engineering, The University of Tokyo.

Original Japanese language edition published by TOTO Shuppan.

＊　　　　　　　　　＊　　　　　　　　　＊

本书中文简体翻译版由 TOTO Shuppan 授权给清华大学出版社在中国境内（不包括中国香港、澳门特别行政区和中国台湾地区）销售发行。

All rights reserved, including the right to reproduce this book or portions thereof in any form without the written permission of the publisher.

Chinese translation rights arranged with TOTO Shuppan, Tokyo through Nippon Shuppan Hanbai Inc.

北京市版权局著作权合同登记号　字图：016523
版权所有，翻印必究。
举报：010-62782989 beiqinquan@tup.tsinghua.edu.cn

责任编辑：汪亚丁
装帧设计：宁成春
责任印制：沈　露

出版发行：清华大学出版社
网　　址：https://www.tup.com.cn，https://www.wqxuetang.com
地　　址：北京清华大学学研大厦A座　　　邮　编：100084
社 总 机：010-83470000　　　　　　　　　邮　购：010-62786544
投稿与读者服务：010-62776969，c-service@tup.tsinghua.edu.cn
质 量 反 馈：010-62772015，zhiliang@tup.tsinghua.edu.cn

印 装 者：涿州汇美亿浓印刷有限公司
经　　销：全国新华书店
开　　本：152mm×228mm　　印　张：12.75　　字　数：207千字
版　　次：2005年12月第1版　　　　　　　印　次：2024年5月第26次印刷
定　　价：55.00元

产品编号：016523-04